Mechanical Design of Instruments

by

P. H. Sydenham
Measurement and Instrumentation Systems Centre
South Australia Institute of Technology

Instrument Society of America
and
The Institute of Measurement and Control

Mechanical Design of Instruments

Printed in the United States of America.

We would like to thank the many suppliers who provided material for this book, and we regret any we may have inadvertently failed to credit for an illustration. On notification we shall insert a correction in any subsequent printings.

INSTRUMENT SOCIETY OF AMERICA
67 Alexander Drive
P.O. Box 12277
Research Triangle Park
North Carolina 27709

Library of Congress Cataloging-in-Publication Data

Sydenham, P. H.
Mechanical design of instruments.
1. Engineering instruments — Design and construction.
I. Title.
TA165.S93 1986 621.8 86-2950
ISBN 0-87664-855-3

Designed by Summit Technical Associates, Inc.

Foreword

As a charter member of the Instrument Society of America, I am personally honored to have the opportunity to welcome the appearance of Professor Sydenham's work. Its publication inaugurates the first of a series of co-publishing ventures by The Institute of Measurement and Control, UK (InstMC) and the Instrument Society of America (ISA). Its publication also reflects the ongoing cooperation between the two organizations, who share aspirations for service to their Members and others in the international technological communities they also serve

As an active member of ISA's *Heritage Group*, I can also welcome this work; it presents aspects of the historical development of instrumentation and, at the same time, emphasizes the current need for an understanding of the important roles such mechanical instruments still play in our professional lives.

The trend toward electronic instrumentation has not entirely replaced the need here and abroad for mechanical expertise, not only in sensing and actuating mechanisms, but also in instrument enclosures and housings. Looking back and looking ahead, one must come to the conclusion that mechanical instruments will be with us for many years to come. Such instruments as bourdon-spring pressure gages, diaphragms, and bellows-operated dial thermometers, including their bimetal versions, are strongly imbedded in our technologies. Laboratories conducting tests and calibrations will continue to use manometric gages for pressure/vacuum standardization. Optical instruments will continue to serve in the laboratory as well as in the field.

We gladly welcome the advances being made in our technologies by electronic and electrical instrumentation, but the contributions made to us by the designers of mechanical instruments and our better appreciation of their ingenuity will enhance our understanding of these vital classes of instruments that have served us so well in the past and will continue to do so in the foreseeable future.

Robert L. Galley, PE
Life Fellow, ISA

Preface

The design of measuring and other fine instruments requires knowledge of many fields of endeavour.

Since the 1950's electronic technique has gradually replaced much of the information processing and calculation that has formerly been done with electrical and mechanical machines.

This trend moved so solidly and rapidly that a generation was brought up to believe that electronic technique can solve most information handling problems.

During the 1960's and 70's this attitude generally sufficed, for then the cost of the electronic part of an instrument constituted a major part of the overall price.

Today the electronic portion of the cost of an instrument system has shrunk to be much less than the mechanical aspect. The time has come for due emphasis to again be given to the design of the mechanical elements that form all instruments.

During a life-long, self-imposed task to produce the literature needed to support improved measurement science and instrument engineering, it became clear that a review of fine mechanics technique was needed to replace a missing gap in the continuum of the available literature. There exists an extensive literature of fine mechanics, but it has just about been buried. It was mostly published before around 1950, an era for which technical libraries now store the books in less-than-easy-to-get-at archives.

To my knowledge no courses on this topic are being taught at this time anywhere in the English speaking world.

It also appears that very little research and development has since taken place into the mechanical aspect of the design of an instrument.

What I perceived to be needed was a reasonably small, copiously illustrated work that ordered the available material in a systematic manner and that directed the user to the detailed, considerable information that is available.

The opportunity to compile such a work arose when I was invited to prepare a series for the journal of the Institute of Measurement and Control, *Measurement and Control,* wherein it was first published.

Comments since received showed that there is a freshening of interest in this aspect of instrument design.

The book is organized to be suited to a teaching course; I have used the material myself, in a condensed form, to form 10 one-hour lectures.

My appreciation is extended to Trevor Warren and the production staff of the Institute of Measurement and Control in London, UK, and to their counterparts at the ISA offices in Research Triangle Park, North Carolina, USA. They each contributed to the presentation of this work.

P.H. Sydenham
Adelaide
March 1986

Contents

Chapter 1
Historical development of technique and understanding

Why revive interest in design of fine mechanism?

The efficient design and application of instrumentation requires expertise in many disciplines. The predominant regimes are electrical/electronic, optical and mechanical. Considerable information is available on electronic method; indeed, it has received so much attention in recent decades that an attitude has arisen suggesting that any instrumentation problem can be solved by persons trained in electronics alone. Topics such as optics and mechanics are no longer treated as fully as they need be in texts and courses on measurement and other instrumentation, with the result that the attitude suggesting that electronics can solve all problems is further enhanced with each generation of trained workers.

The time has come to begin a revival of interest in the mechanical design aspects of instruments. There is need to restate the skills and procedures won in the past as they are in danger of being lost. This may seem to be an exaggeration but it is not. There are very few books in print today that lay out the fundamentals of mechanical design specifically for instruments and these are written as reference sources, not as educational texts that lead the reader onward with increasing depth of understanding of a systematic approach to design. To this must be added the fact that relevant courses are virtually non-existent in English-speaking educational institutions and that there are few specific research groups in such institutions whose role is to advance and disseminate mechanical instrument design knowledge. In contrast, the European continental countries still maintain strong interest in this aspect of instrument design.

This lack of organized effort probably has occurred because of the erroneous belief that the day of the mechanical instrument has passed. There has been a decline in need, but there no longer exists the same proportionate offering in books and courses as there was in, say, the 1950's, when mechanical computing apparatus was nearing a peak of maturity and application. At that time many high level courses and sophisticated books were created to fill the then fast emerging need for machine computation. The arrival of electronic alternatives very rapidly displaced mechanical methods and this began a trend away from mechanical solutions. This trend was so dominant that it created a firm impression that the mechanical design aspects of information machines were no longer of any real relevance and that they need no longer be taught or researched.

This swing away has been overemphasized with the result that action is now needed to restore a more reasonable balance of interest in mechanical design aspects of instruments. Fundamental training and knowledge in mechanical design is still as necessary as it has ever been.

A first reason for taking greater interest is that instrumentation sub-assembly costs are rapidly changing so that the mechanical proportion is now appearing as a most significant cost component, this being due to the enormous reductions that have occurred in the cost of the electronic assemblies that formerly were a significant cost factor. Instruments are generally designed to a selling price; they are not sold nor necessarily costed on the degree of internal sophistication, of which the user is often not aware. That an instrument contains thousands of active solid-state elements arranged in most complex and sophisticated assembly may, at some time, create an air of awe in the user, but the critical use factor is basically that the device performs the task expected at the best price. In contrast, the mechanical aspects, while being far less complex and simpler in concept that electronic assemblies, are vital to the operation (Figure 1.1). Their cost is now becoming a very significant factor. It is not hard to see that instrument design will

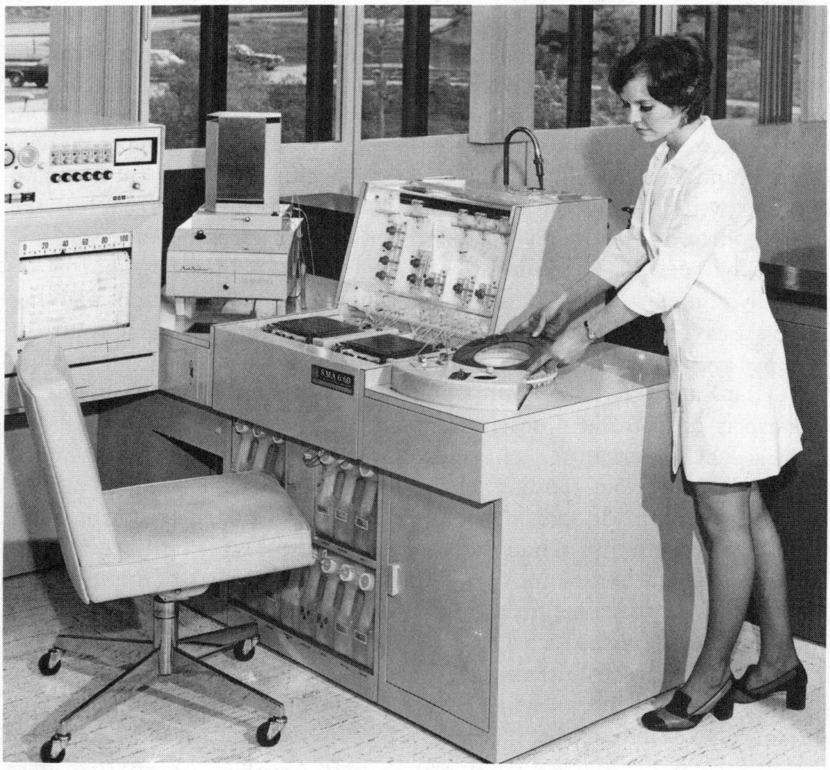

Figure 1.1 This multi-parameter chemical analysis equipment is one example of the many instruments systems that use fine mechanisms. (Courtesy Technicon Equipment Pty. Ltd, NSW.)

now tend to concentrate on the mechanical regime in order to seek replacement by integrable electronic alternatives; where this is not possible, better ways will be sought to provide the necessary mechanical function in an improved manner.

Although many of the traditional mechanical functions used in sensors and actuators can, and will, be replaced in this way, the fact will remain that the world is substantially composed of mechanical structures and mechanisms and that man's attempts to create his own kind of systems always need mechanical elements. Even semiconductor devices rely on fine mechanics for their design and manufacture and coherent construction.

A second reason for increasing the attention given to mechanical design aspects of instruments is that modern developments in instrument design — through the use of mathematical and, hence, computer modelling of the various classes by generalized models — is based upon the fact that many different regimes, such as mechanical, electrical and acoustic, can each be described by common mathematical models. This rigorous sytems approach to the understanding of the static and dynamic regimes of instrument systems requires good grounding on mechanical components, one that is more specifically tailored to information mechanisms than is the content of currently available works on the design of power transmission mechanical systems.

A third reason is that, as instruments will continue to be made with some proportion of mechanical components, their users must be aware of the characteristics of mechanical elements in order to obtain efficient operation and worthwhile service programs.

A trend is currently emerging toward the design and use of complex multidisciplinary electromechanical systems that, while having the fineness of instruments, are not generally regarded in this light — this is occurring in the field of robotic mechanisms, for example. Now that designers can give such systems a really useful degree of 'intelligence' because of the great cost and size reductions that have come with electronic data processing, designers are in a position to provide very useful, general-purpose robotic devices. Electronic methods are well in hand and are continuing to be developed extensively. Mechanical skills, however, now need to be enhanced to match the high level of performance now possible from the sensors and actuators of some robotic machines.

Related areas, such as prosthetics and highly precise and complex medical operating apparatus are also pressing for improved mechanical design and production.

These chapters have been written to assist those who are feeling this pinch. They review what is known as a broad statement and state where that knowledge is recorded, for much is now out of print but still very applicable. This is accomplished by first reviewing the historical progress of the mechanical aspect of instrument design and then by concentrating, in turn, on several aspects which have emerged as the accepted sub-sets of the relevant knowledge. Throughout the text the material presented is supplemented by illustrations of practice from the particular period of history or from whichever discipline was in vogue. References are given to key works that have been published in which useful information can be found. Many of these works are not in print for, as already stated, it is rarely possible to provide current, in-print sources due to the decline in interest in the aspect of instrument design. Many of the earlier works will find new relevance as current users begin to appreciate the kind of knowledge that has been presented in the past. Although many of the early texts are generally not available in the average local library, they are normally held in major national collections and are freely available through interlibrary loan services. Contrary to a misguided popular belief that works of technological nature that are over a decade or two in age are of little current value, it will be seen that those on mechanical design have not lost their relevance to the same degree as has occurred in some other areas.

It is the intention of this text to provide a general source that will enable the reader to find and use the considerable body of knowledge that has been accumulated in the past on mechanical aspects of instrument design, and in doing so provide an awareness of the historical development that will give confidence and appreciation of proven practice to those engaged in this area.

Mechanical design over the ages

Instrumentation has its genesis as part of the daily needs of the earliest settled races who found a need to regulate their new found gregarious life style in such ways as timekeeping, weights and measures, and astronomical observation. Each of these activities required the conversion of natural resources into implements of measurement and control, using their intuition and gradually accumulated experience.

The first known devices were mechanical in nature and took the form of weighing balances of very simple design. The oldest existing today dates from around 4500 BC; it was found at Naquada in Egypt.

Balances, like all measuring instruments, were gradually improved in sensitivity and convenience as the centuries passed. By the Roman times they were much more advanced in terms of the mechanical principle involved, but they were nevertheless still very basic. While improvement was achieved and more sophistication was developed, it is important to recognize that even the simplest mechanisms and ideas have been retained where they match the need.

In other fields mechanical ingenuity was put to use to devise and construct such instruments as forms of clocks, water dispensers, planetarial devices, south-seeking chariots, and hodometers for recording distances travelled during long marches (Sydenham, 1979).

Possibly the most remarkable of these mechanical devices is an object known today as the Antikythera mechanism (from the place where it was found in the sea in AD 1900) (see Figure 1.2). This corroded lump has slowly yielded its structure which is now known to be a quite complex calendrical computing mechanism containing dozens of gears. It uses epicyclic and differential gear arrangements. The complexity of this mechanism is so far advanced compared with other extant objects and recorded knowledge of that time that it seems to be quite out of place in the historical development of ancient mechanical engineering achievement (De Solla Price, 1974).

Although the ancients recorded details of many ingenious mechanisms, almost none has survived as artifacts. Furthermore, the detail now available generally does not provide enough design information to enable unique reconstructions to be made. Much of what we know about mechanical devices of those times is through the reconstructions of past historians, sometimes over-enthusiastic and many with very little engineering training.

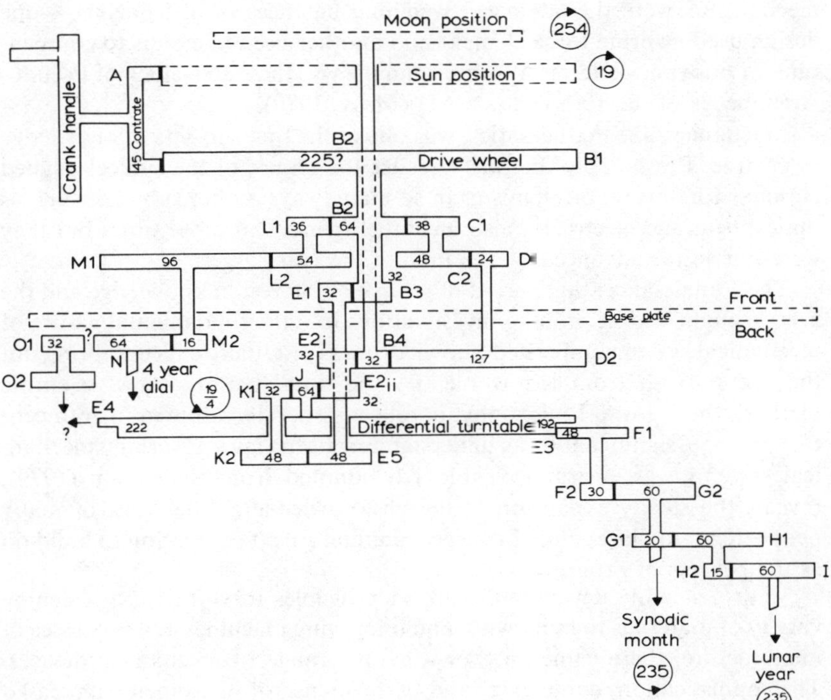

Figure 1.2 Gearing of the Antikythera mechanism of ca 80 BC (De Solla Price, 1974) (©American Philosophical Society. Used with permission.)

Only a fraction of the recorded knowledge of the ancient times has survived. Much of that which does exist has come to us through many translation stages from one language to another. Certain works, such as those from the Aristotelian School of the Fourth Century BC, enunciated the basic laws of levers, and from such works it is clear that the thinkers of the times were beginning to seek principles of operation of mechanical systems rather than relying solely on the intuitive designs that craftsmen obtained through experience. A good introduction to the historical development of technology, which was predominantly mechanical in the ancient times, can be found in Burstall (1970).

The arrival of, and reasons for, the Dark Ages period (beginning around AD 200) halted the previous steady progress of the development of the technologies at a time when Alexandrian science was apparently ready to begin a significant stage of development in practical application. Recorded knowledge that has survived then passed over to the Eastern nations who added to it, adapted it, but often distorted it. During the Dark Ages period in Europe the Arabians advanced many facets of knowledge including the design of quite complex and sophisticated mechanical devices. Notable mechanisms were the advanced weighing balances of al-Khazini — one design used a spring force element and another used five pans to compensate for buoyancy effects. Another example was the water-clock of Pseudo-Archimedes of the 10th century AD (Mayr, 1970).

Mechanics, like mathematics, was one of the first sub-sets of knowledge to emerge. Presumably this was because the nature of the subject aligned closely with natural mechanisms in so many ways. Other subjects, such as optics, heat and electricity, also have their genesis in those times but they were not so far advanced by the time of the Dark Ages.

The Renaissance, that period of rebirth of interest in knowledge and the arts, brought with it probing investigations into the fundamental nature of mechanical systems. Inherited knowledge was gradually culled to bring out the useful truth from the whole. By the time that the 'new' scientific method, the Natural Philosophy, had been generally adopted in 16th century Europe, significant basic understanding of the laws governing mechanical systems was emerging. Table 1.1, adapted from Sydenham (1979), reveals the steady expansion of design knowledge as the work of many people fell into place, piece by piece, enabling a next generation to build on as the process of progress.

With a steadily developing bank of principles to work to, a widening variety of materials to build with, and improving machines and processes of manufacture, there came an ever-widening range of mechanical devices. The sophistication, complexity and performance of instruments increased steadily over time.

Clocks and clock-work mechanism requirements provided strong incentive to improve mechanical design. There is evidence suggesting that the Chinese provided the bridging link through the Dark Ages between

Table 1.1
Introduction of Mechanisms and Theories of Mechanical Design

Ancient Times

Mechanisms have developed by intuition since the earliest times. Instrument mechanisms grew alongside power devices as a much smaller indistinguishable group. Mechanics, as a science, has its roots in the earliest times. By the 1400's, mechanisms and physical principles known (to some level) included the following. They were not known by all peoples simultaneously at any given time.

Mechanisms	*Fundamental Theories*
Rotary to linear motion	Strengths of materials in compression, tension and bending
Linear to rotary motion	Laws of levers
Lever balance	Mechanical advantage of multiple pulley system
Toothed gear wheel	Regulator
Wheel	Specific gravity
Wedge	Buoyancy
Pulley	Center of gravity
Bearings and lubrication	Equilibrium of floating bodies
Locks with key	Friction
Screw thread	
Drawing compass styles	
Wormwheel and worm	
Differential gear train	
Epicyclic gear train	
Ball bearings	
Gimbals	
Graduated scales	
Various escapements	

Post Dark Ages

	A.D.
Applied mechanics and strength of materials begun as science (Galileo)	c 1600
Equilibrium of forces (Stevin)	c 1610
Laws of orbiting bodies (Kepler)	c 1615
Hydrostatic pressure exerted equally (Pascal)	1647
Huygens constructed first pendulum regulated clock	1656
Laws of elasticity (Hooke)	c 1665
Theory of compound pendulum (Huygens)	1673
Laws of motion, gravitation (Newton, the 'Principia')	1687
Theory of columns in compression (Euler)	c 1750
Atwood's 'On the rectilinear motion and rotation of bodies'	1784
Modulus of elasticity (Young)	1807
Vibrations of strings, rods, plates (Chladni)	c 1809
Theory of beams for wood (Dupin)	1815

Table 1.1 (Cont'd)

Theory of beams for metals (Duleau)	1820
Elastic limit as design basis (Navier text)	1826
Internal motion of particles in fluid (Brown)	1827
Poisson ratio (Poisson)	1829
General equations asembled for strength design (Lamé and Clapeyron)	1833
Kinematics terms introduced (Ampere)	1834
Recognition of stress raisers as cause of fatigue failure (Rankine)	1843
Autocontrol principles begin to emerge in steam engine systems	c 1850
Gyroscope invented (Foucault)	1852
Analysis of continuous beams (Clapeyron)	1857
Foundations laid of metallurgical science	1860s
Combinations of stresses in materials (Mohr)	1868
Strain energy used in structural analysis (Castigliano)	1873
Systematic foundation of mechanism laid (Reuleaux's 'Kinematics of machines')	1875
Stress of bodies in contact (Hertz)	c 1880
Theory of lubrication (Reynolds)	1886
Specialist texts become prolific	post 1900
Brownian motion research (Perrin; Millikan; Fletcher and others)	c 1905
Creep in metals under stress and elevated temperature (Andrade)	1910
Creep knowledge reviewed (Bailey)	1935
Metrology established as subject in Production Engineering	c 1940

Note: For further detail consult Burstall (1970) and Sydenham (1979).

mechanical skills of the Hellenistic era and the Western European world. Their contribution, however, did not continue much past AD 1100 — see Needham's works on Chinese science and technology.

Clock-work mechanisms were crudely made in the 14th century (Figure 1.3) but skills, experience and materials improved greatly as time passed. Harrison's time-keepers, on show at the Maritime Museum, Greenwich, England show the sophisticated level of design and manufacturing skills that were available by the mid-18th century.

Clock-work stimulated interest in automation mechanism, and since the 16th century AD there have been made all manner of incredibly complex mechanical devices. Whole automation orchestras were made to play set pieces. Mechanical animals, birds and insects were created that performed quite remarkable dynamic functions. But these devices were built as the result of craft; they were based upon very little theoretical foundation.

Figure 1.3 Dover Castle clock built in the 14th century (Ward, 1973) (©Science Museum, London, UK. Used with permission.)

Much of the early instrumentation that was produced by trade houses of the 16th to 19th centuries was built by mathematical instrument makers who were also greatly concerned in those times with the appearance of their products. This was an important specification, for they were seeking to sustain considerable sales into the drawing-rooms of the wealthy who patronized the workshops in search of objects of amusement.

In the 1770's Ramsden built circle-dividing engines that were able to divide circular scales to a previously unobtainable precision with greater reproducibility than had been seen before. He was able to produce surveying instruments of high and consistent quality and performance. This, and other events of similar character, steadily led to the concept of mass production on the basis of interchangeable parts, this concept being first introduced into practice in the 19th century for the manufacture of firearms and clocks.

The Victorian era saw a considerable increase in electrical knowledge that provided basic principles for the design and manufacture of many varied electromechanical devices. Figure 1.4 shows detail of an arc lamp. The galvanometer was continously improved in sensitivity to a point whereby 1900 Brownian motion of the mechanical components was the limiting parameter on detection sensitivity.

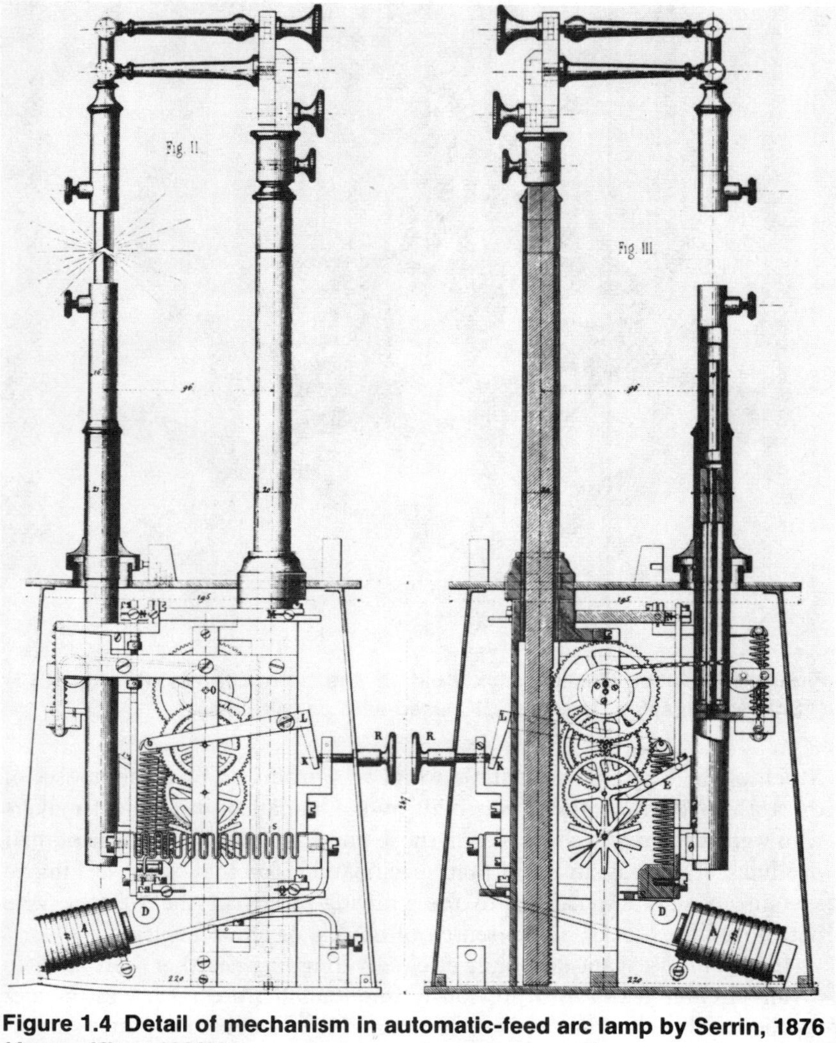

Figure 1.4 Detail of mechanism in automatic-feed arc lamp by Serrin, 1876 (James-King, 1962)

The stored-program computer concept was first implemented in the early 19th century in the mechanical calculating engines of Babbage. Unfortunately, production skills of that time did not match the design requirement for high precision of manufacture of its components and the machines were not successful.

It was in the 19th century that the theoretical mechanics of linkages came to a stage of maturity in which sophisticated mathematics was developed in the search for such elusive mechanisms as the exact straight-line linkage. Out of this work came contributions like the Chebyshev functions now used so generally in electronic filter realization. Papers by Ferguson (1962, 1977) provide insight into the historical development of mechanism theory and pictorial representation of mechanism, respectively.

Published texts of relevance

At the start of this century a considerable body of knowledge existed that enabled mechanical design aspects of instruments and other fine mechanisms to be carried out in a systematic manner. There began to appear many works that were more fundamental in nature. These works supplemented the many volumes that had been published since the earliest days of printing on the kinds of mechanisms that could be be built. These catalogs of ideas played a useful part in the development of mechanical design. When combined with works of a theoretical nature, design could become a more rigorous and efficient procedure of converting resources into tools of information processing.

Published works about types of ingenious mechanisms were numerous, a selection being Barber (1906), Barlow and Herschel (1848), Drake (1879), and Robison (1882).

Collections of mechanical instruments, made as small distinct assemblies that embodied principles for demonstration purposes, were held by many teaching institutions. Those of the University of Utrecht Museum (Bos, 1968) and the Van Marum collection at Haarlem (Turner, 1973) are fine examples still existing. Christopher Polhem, in 1729, systematically designed a 'mechanical alphabet' in which he grouped known mechanisms as the 'vowels' and 'consonants' of his alphabet. Some of his wooden models (Figure 1.5), have survived and are held in the Swedish National Museum of Science and Technology, Stockholm. Use of model collections is no longer so noticeable in teaching institutions.

In the closing decades of the 19th century there began to appear, first as parts of works and then later as complete texts, works laying out the design fundamentals of mechanical instrumentation apparatus. Thompson and Tait (1883) contains material suggesting this trend. Glazebrook (1922) has a small chapter in Volume III, by Horace Darwin and Cecil Mason, that is concerned specifically with the design of scientific instruments, the emphasis being on the mechanical aspects. Strong (19[illegible]8) and Jacobs (1943) also

Figure 1.5 Some of Polhem's models from his "Mechanical Alphabet" (National Museum of Science and Technology, Stockholm)

contain chapters on mechanical aspects of design. Specific works on mechanism designs, for either power or instrument applications, mostly appeared from the 1930's onwards, presumably as the result of a considerable upsurge in interest in production and use of mechanical computing mechanisms (Figure 1.6) and a general upturn in needs for economically designed instruments and other mechanical products. Example texts are Whitehead (1934), Beggs (1955), Braddick (1954), Geary (1954, 1955, 1960, 1961, 1962), Mabie and Ocvirk (1958), Shigley (1963), Berg (1965), Andreeva (1966), Martin (1969), Lent (1970), Trylinski (1971), Davidson (1974), Deutschman *et al.* (1975), Hughes and Martin (1977). Note that few of these are concerned specifically with instrument mechanics.

The mechanical design aspect of instruments, however, can also be studied by consulting applications-based texts which discuss apparatus involving a significant degree of fine mechanism. Stanley (1890) includes an

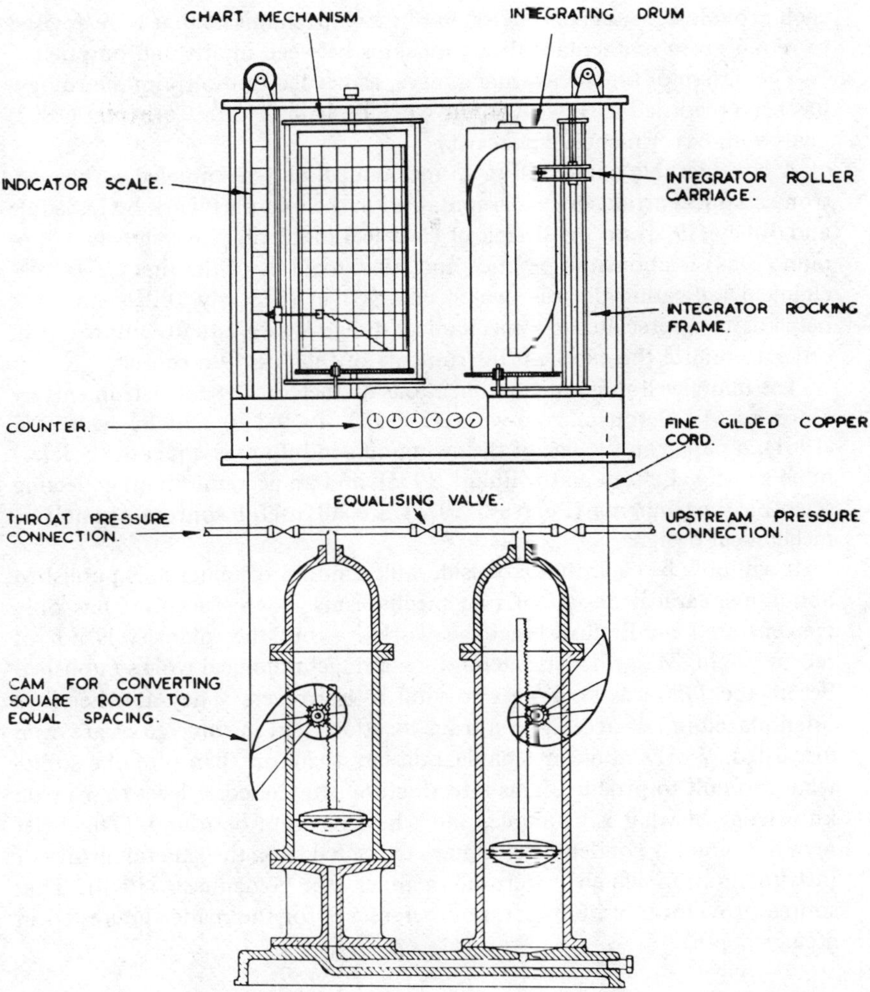

Figure 1.6 Mechanisms involving mechanical computing found increased use in the first half of the 20th century. Combined flow recorder by George Kent, ca 1900 (Ardley, 1953) (©Brown Boveris Kent. Used with permission.)

introductory chapter about the mechanical manufacture of surveying instruments. Chemical balance design is the subject of Walker (1887), Brauer (1909), Stock (1969) and Wolsky and Zdanuk (1980). Mechanical computing mechanisms, which first appeared in the Greek times as the Antikythera mechanism and other planetarial mechanisms, became subjects of more rigorous and systematic design in the late 19th century. Thomson and Tait (1883), International Hydrographic Bureau (1926), Murray (1948), Svoboda (1948), Eames and Eames (1973), and Pugh (1975) each provide accounts of the design of these mechanisms that were devised to retain given mathematical relationships between inputs and outputs.

The design of fine measuring gauges, as needed in toolroom metrology, has been recorded in a two-volume work by Rolt (1929). Carpenter (1892) deals with experimental engineering.

A considerable contribution to mechanical design knowledge for electromechanical arrangements is contained in the two-part work by Drysdale and Jolley (1924) on the design of electrical indicating instruments. Chipman (1964) is about the practice and fundamentals of the first electromechanical indicating devices, which emerged in the early 19th century: it helps us to appreciate how mechanical design was gradually improved in order to realize the excellent instruments of the late 19th century.

The many well-illustrated publications on meteorological instruments by Knowles-Middleton and co-workers (1953, 1969a, b) and by Multhauf (1961), a paper on the use of the swinging pendulum as a means to determine gravity (Lenzen and Multauf, 1965), and an account about collecting scientific instruments (Pearsall, 1974) are all useful sources about fine mechanical design.

It will now be clear that a considerable amount of material is published about mechanical design of fine mechanisms — this account has only concentrated on English-language works — but the majority is not of recent origin. Major libraries tend to regard technological works published before the 1950's as no longer so vital to borrowers. They are therefore often placed in less accessible storage locations and in some cases are even discarded. Works about mechanical design, therefore, can often be somewhat difficult to procure. This introduction is intended at least to provide knowledge of what is available, and where it might be found. (This brief review is a highly condensed summary extracted from the general history of instrumentation which covers all regimes, see Sydenham (1979). That source provides more bibliographic references for the reader interested in greater depth).

References

Andreeva, L. E. 1966. *Elastic elements of instruments*. Israel Program for Scientific Translations, Jerusalem.

Ardley, W. G. 1953. 'The history of flow measurement by differential pressure. IV. The development of Venturi Recorders from 1894 to 1911', *Instrument Engineer*, **1** (4), 83–86.

Barber, T. W. 1906. *The engineer's sketch book of mechanical movements, devices, appliances, contrivances and details*. Spon, London.

Barlow, P. and Herschel, J. F. W. 1848. *The encyclopaedia of mechanical philosophy*. John Joseph Griffin and Co, London (also Richard Griffin, Glasgow, 1854).

Beggs, J. S. 1955. *Mechanism*. McGraw-Hill, New York.

Berg, W. M. 1965. *The theory and application of precision mechanical components and a guide for engineers, designers, draftsmen and technicians*, 2nd edn. Sacken, East Rockaway.

Bos, H. J. M. 1968. *Mechanical instruments in the Utrecht University Museum*. Utrecht University Museum, Utrecht.

Braddick, H. J. J. 1954. *The physics of experimental method*, Chapman and Hall, London.

Brauer, E. 1909. *The construction of the balance*. Incorporated Society of Inspectors of Weights and Measures, London.

Burstall, A. F. 1970. *A history of mechanical engineering*. Faber and Faber, London.

Carpenter, R. C. 1892. *A text book of experimental engineering for engineers and students in engineering laboratories*. Wiley, New York.

Chipman, R. A. 1964. 'The earliest electro-magnetic instruments'. Paper 38, 21–26, United States National Museum, Bulletin 240: Smithsonian Institution, Washington, DC.

Davidson, A. 1974. *Handbook of precision engineering*. 7 vol. to 1974. Macmillan, London.

De Solla Price, D. 1974. 'Gears from the Greeks: the Antikythera mechanism — a calendar computer from ca 80 BC'. *Trans Am Phil Soc*, **64** (7), 3–70. Also from Science History Publications, New York.

Deutschman, A. D., Michels, W. J. and Wilson, C. E. 1975. *Machine design theory and practice*. Macmillan, London.

Drake, J. A. 1879. *The practical mechanic: comprising a clear exposition of the principles and practice of mechanism, with their application to the industrial arts*. J. W. Lukenbach, Philadelphia.

Drysale, C. V. and Jolley, A. C. 1924. *Electrical measuring instruments, 2 parts*. Ernest Benn, London.

Eames, C. and Eames, R. 1973. *A computer perspective*. Harvard University Press, Massachusetts.

Ferguson, E. S. 1962. 'Kinematics of mechanisms from the time of Watt'. Paper 27, 185–230. United States National Museum, Bulletin 228. Smithsonian Institution, Washington, DC.

Ferguson, E. S. 1977. 'The mind's eye: nonverbal thought in technology'. *Science*, **197** (4306), 827–36.

Geary, P. J. 1954. *Flexure devices — Pivots, movements, suspensions*. Sira Institute, Chislehurst, Kent, UK.

Geary, P. J. 1955. *Knife edge bearings* (Survey of instrument parts — No 2). Sira Institute, Chislehurst, Kent, UK.

Geary, P. J. 1960. *Torsion Devices* (Survey of instrument parts — No 3). Sira Institute, Chislehurst, Kent, UK.

Geary, P. J. 1961. *Instrument ball bearings* (Survey of instrument parts —No 4). Sira Institute, Chislehurst, Kent, UK.

Geary, P. J. 1962. *Fluid film bearings* (Survey of instrument parts — No 5). Sira Institute, Chislehurst, Kent, UK.

Glazebrook, R. T. 1922. *A dictionary of applied physics*. Macmillan, London (Vol 1, 2 — 1922; Vol 3-5 — 1923).

Hughes, J. H. and Martin, K. E. 1977. *Basic engineering mechanics*. Macmillan, London.

International Hydrographic Bureau 1926. 'Tide predicting machines'. Special Publication No 13, July.

Jacobs, D. H. 1943. *Fundamentals of optical engineering*. McGraw-Hill, New York.

James-King, W. 1962. 'The development of electrical technology: 3. The early arc light and generator'. Paper 30, 333–407, United States Museum, Bulletin 228, Smithsonian Institute, Washington, DC.

Knowles-Middleton, W. E. and Spilhaus, A. F. 1953. *Meteorological instruments*. University of Toronto Press (1st edn — 1941).

Knowles-Middleton, W. E. 1969a. 'Catalog of meteorological instruments in the Museum of History and Technology'. Smithsonian Institution Press, Washington, DC.

Knowles-Middleton, W. E. 1969b. *Invention of the meteorological instruments*. Johns Hopkins Press, Baltimore.

Lent, D. 1970. *Analysis and design of mechanisms*. Prentice Hall, Englewood Cliffs.

Lenzen, V. F. and Multhauf, R. P. 1965. 'Development of gravity pendulums in the 19th century'. Paper 44, 301–348, United States National Museum, Bulletin 240. Smithsonian Institution, Washington, DC.

Mabie, H. H. and Ocvirk, F. W. 1958. *Mechanisms and dynamics of machinery*. Wiley, New York.

Martin, G. H. 1969. *Kinematic and dynamics of machines*. McGraw-Hill, New York.

Mayr, O. 1970. *The origins of feedback control*. MIT Press, USA.

Multhauf, R. P. 1961. 'The introduction of self-registering meteorological instruments'. Paper 23, 95–116, United States National Museum, Bulletin 228. Smithsonian Institution, Washington, DC.

Murray, F. J. 1948. *The theory of mathematical machines*. Kings Crown, New York.

Pearsall, R. 1974. *Collecting and restoring scientific instruments*. David and Charles, Newton Abbot.

Pugh, J. 1975. *Calculating machines and instruments*. Science Museum, London (1st edn — 1926).

Robison, J. 1822. *A system of mechanical philosphy*, 4 vols. John Murray, London.

Rolt, R. H. 1929. *Gauges and fine measurements*, 2 vols. Macmillan, London.

Shigley, J. E. 1963. *Mechanical engineering design*. McGraw-Hill, New York.

Stanley, W. F. 1890. *Surveying and leveling instruments*. E & F Spon, London.

Stock, J. T. 1969. *Development of the chemical balance*. HMSO, London.

Strong, J. 1938. *Modern physical laboratory practice*. Blackie and Son, Glasgow.

Svoboda, A. 1948. *Computing mechanisms and linkages*. McGraw-Hill, New York (MIT Radiation Laboratory Series).

Sydenham, P H. 1979. *Measuring instruments — tools of knowledge and control*. Peter Peregrinus, London.

Thomson, W. and Tait, P. G. 1883. *Treatise on natural philosophy*. Cambridge University Press, England (many versions were printed).

Turner, G. L'E. 1973. *Van Marum's scientific instruments in Teyler's Museum*. Noordhoff Int Pub, Leyden.

Trylinski, W. 1971. *Fine mechanisms and precision instruments — principles of design*. Pergamon Press, Oxford, translated by A. Voellnagel from Polish original.

Walker, J. 1887. *The theory and use of the physical balance*. Clarendon Press, Oxford.

Ward, F. A. B. 1973. *Clocks and watches, 1: Weight driven clocks*. HMSO, London.

Whitehead, T. N. 1934. *Instruments and accurate mechanism — underlying principles*. Macmillan, London (also Dover Pubs, New York, 1954).

Wolsky, S. P. and Zdanuk, E. J. 1980. *Ultramicroweight determination in controlled environments*. Wiley, New York.

Chapter 2
Materials — working with imperfect resources

The importance of materials

Materials are the many kinds of matter from which objects are formed. In instrument systems they are used in numerous forms for numerous reasons. They might, for instance, be used to provide a structural framework that holds other components in some fixed or allowably moveable juxtaposition. Alternatively, they may be chosen because of the possession of a property concerned with energy conversion that makes them suitable as a transducer element, a piezoelectric crystal being an example.

The choice of a particular material may rest on the constancy of some of its properties being present while others vary in a predictable and comparatively large manner. In other applications the designer seeks a material that has only satisfactorily constant properties. Each application requires careful assessment of the total system needs, these then being matched as closely as possible to those of an available substance.

Just where the use of a material ceases to be a mechanical one and becomes one of another regime, such as electrical, optical, and the like, is often hard to define. Because materials form a semiconductor circuit in an integrated circuit does not imply that knowledge of mechanical factors is of negligible importance. In practice the mechanical properties of the material must be well known and controlled if the desired electronic characteristics are to be obtained in the production of integrated circuits. In choosing material for a component part of an instrument system, recognition must be given of the prime and secondary functions of the component.

Centuries of experience with the common materials such as iron, brasses, copper, tin, and others has proven their worth to such an extent that designers instinctively know reasonably accurately which one is right for which task. This skill is especially well developed for the design of the heavier, non-instrument kinds of structure, for which training is well developed and the literature is well spread and read.

Highly developed instrument design, however, often dictates the additional use of quite exotic, hardly known, highly controlled materials in relatively small quantities. General knowledge of these is very poor, and as a rule it is only a person deeply involved in a specific user branch of instrument production who will know of some of them. This chapter concentrates more on the unusual than the usual, it being assumed that users of this text will be generally knowledgeable about the more common engineering materials.

The instrument designer will often be unable to find a material that possesses a sufficient number of perfect enough parameters. In most cases it is necessary to make the best compromise that can be found, compensating for deficiencies by various strategies. New materials are being developed continously, but in instrument work the demand for a new special-purpose alloy or other substance is often not large enough to encourage the research and commercial manufacture that might overcome the limitation. Throughout history, materials have been devised that made considerable impact in the design of instruments — Invar, the alloy with a very low thermal coefficient of expansion, being possibly the most notable. In general, however, materials have first been made available for the power form of machine, instruments taking them up as they became available.

The success or otherwise of an instrument may rest heavily on the designer's use of materials. Careful selection can often remove the need for compensation, reduce the operational limitations, lower the overall cost, and considerably enhance performance. Any instrument systems design procedure must recognize that the materials used are as vital as the principles and systems arrangement employed. It rarely pays to skimp on the choice to perhaps reduce the cost, for instrument systems are designed to make beneficial use of the specific properties of common and unusual materials. Once the required specifications for the component are decided from the functional needs, it will be found that the range of choice of materials will have narrowed considerably. In many cases the designer reaches the point where one material alone comes close to (but may not reach) the requirement. The only decision then is which supplier to use; there may be only one or two manufacturers in the whole world.

Instrument applications often show up the unexpected, especially those involving performance at the frontier of experimental knowledge. For example, materials thought to be non-magnetic may be found to possess an unwanted degree of remanent magnetism that was not observed before or reported in the product data sheet; so-called zero thermal expansion materials are found to exhibit some degree of thermal expansion or contraction when the degree of thermal stability is controlled enough. The designer must constantly be on guard to ensure that some unwanted 'second-order' effect has not been unwittingly built in — such as the Hall effect in the use of copper.

It is important to realize that the properties of materials are numerical numbers assigned by man to the material as the result of actual experimental observations based upon the prime physical standards of measurement, which also have inherent uncertainty. Many quoted coefficients hold only for given conditions of such parameters as temperature, degree of mechanical working, and other 'history' variables.

Obtaining data on material properties

Unlike the situation existing in heavier mechanical engineering design where the materials used are well documented and widely used, data on the more unusual kinds that are needed in an instrument design will often be very difficult to procure. All too often the parameter that is needed is not given in the already long list of properties provided by the maker of the material. Those constants quoted have been gained after considerable experimental effort by the maker; naturally the properties that are measured reflect the design needs of the bulk of users for that material. Information on properties often has to be obtained by patient search of the literature and by obtaining in-house publications issued on the specific materials by their makers. It seems there is no suitable extensive handbook that includes all the properties that instrument designers often need. Many scarcely rate a mention, for example, the thermoelastic coefficient and the degree of mechanical hysteresis.

The international computer-based data bank, CODATA, is an obvious choice when seeking less publicized properties. Published standards — such as ASTM, US Military, SAEME, AFNOR, BS, AECMA, DIN, ASME — exist on many materials. These define what the material should be and are useful guides to current practical achievement in materials production.

As will be seen below, the possible parameters that could be listed for any given product run into well over 50 so-called constants. Many of these constants are, in reality, variables depending upon other parameters, few of which have been satisfactorily described by mathematical expressions. The task of compiling a handbook of material properties for instrument designers does not appear to have attracted the attention of an author as yet.

Standard engineering design texts and handbooks include the more common data but invariably lack the other necessary, less common numerical values needed in instrument design. Examples are Faires (1955) and Titcomb (1966). Several extensive works are in print on physical constants. Of the many books published in English on instrument design over the past decades, it appears that only Braddick (1954) and Trylinski (1971) have included a substantial account of the materials used in general instrument construction, but neither takes the reader much beyond an introduction. Texts on what has become known as materials science are valuable for their explanations of the physics and chemistry of materials, but they, too, generally lack hard data of the special-purpose materials.

In-depth data could be expected and sometimes is found in more specific works. Strong (1938) for example, contains a chapter on the use of fused silica and one on the 'materials of research'. Drysdale and Jolley (1924), part 1, has an extensive chapter on electrical materials, which is slanted toward instrument needs. Copper and its derivative alloys, such as beryllium copper, phosphor-bronze, monel, nickel-steel and many more, are the subject of a book published by the Copper Development Association (CDA, 1953). This book is a fine example of the kind of text that the instrument designer would like to have about the many other materials that are used. This Association has published several works on the use of copper. An in-house booklet by Brush Clevite (1966) describes piezoelectricity and the materials involved. Optical materials, by contrast, are well covered in the numerous texts dealing with the infrared, visible and ultraviolet regions. These directly relate to instrument design problems because their general bias is toward scientific optics.

It is most important to realize that material properties depend greatly on composition, heat treatment, and cold working (Figure 2.1). Each batch can vary greatly, and care must be exercised in securing material with a known history if reprocessing is to be avoided.

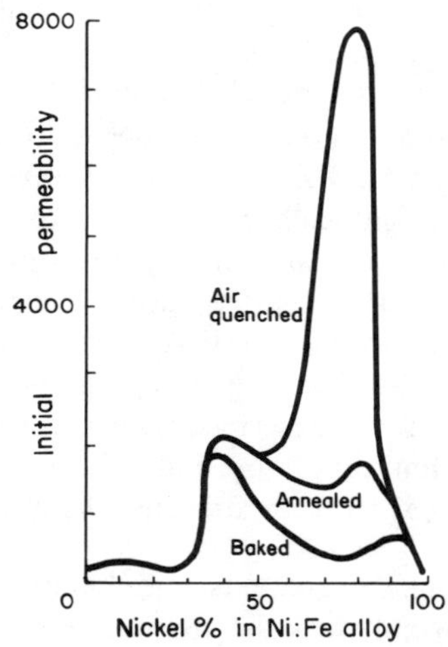

Figure 2.1 Material properties depend greatly on composition, heat treatment, and cold working as shown by this graph of initial permeability of nickel-iron alloys (Wiggin) (©INCO EUROPE Ltd. Used with permission.)

Parameters and Properties of materials . . . a checklist

This section gives the terminology and brief definitions of the majority of materials parameters that are encountered in instrument design plus some other factors that may need consideration. It is not absolute — no list ever could be. A material will generally be specified in terms of a selection from it, but never all. Many will be of no obvious practical relevance to some materials and for this reason will not be available. They are grouped according to the classifications usually used in manufacturer's data sheets and in handbooks. Actual numerical values must be sought by reference to data sources, remembering that they are often dependent on the magnitudes of other related variables. References are given to those parameters not so easily found explained in general engineering texts. The list is intended to be a checklist and source of quick reference.

A. Mechanical and Physical

Pertaining to strength

Tensile or ultimate strength, tenacity — maximum load in tension that can be withstood; usually expressed as stress, not absolute load.
Compressive strength — similar to above but for compression.
Ductility — percentage of elongation after tensile fracture has taken place.
Brittleness — lack of ductility, especially in shock situtations; opposite of toughness.
Malleability — similar to ductility.
Endurance or fatigue limit — maximum stress allowable for unlimited stress cycling without failure.
Toughness — ability to absorb energy after impact.

Pertaining to elastic properties

Modulus of elasticity, Young's modulus — measure of elasticity, stiffness, springiness of a material; is property of material, not its condition of working. Reciprocal effect is compliance of a member. Not to be confused with modulus of a spring, which is its rate of deflection per unit force.
Shear modulus, transverse modulus of elasticity, modulus of elasticity in shear — elasticity of member in shear stress, rather than tensile stress.
Poisson's ratio — ratio of change of transverse dimension to longitudinal change in dimension during elastic strain.
Elastic drift, recovery, relaxation, after-effect, retardation — change of strain that follows after stress becomes stable; may take seconds to months to occur; related to elastic hysteresis.
Elastic limit, limit of proportionality — maximum strain excursion that will return to zero strain without plastic permanent deformation.
Thermal (or temperature) coefficient of elasticity thermoelastic coefficient, thermal coefficient of Young's modulus variation, TEC — reduction (usually of this sign) in the elasticity as temperature rises. Some special alloys have opposite sign coefficient. (Trylinski, 1971; Wiggin, 1972; Huntington, 1975; CDA, 1953; Hamilton, 1975; Vacuumschmelze, 1977).
Hardness — ability of surface to resist indentation, relates to wear; expressed in Mohs, Micro, Knoop, Rockwell, Brinell, Shore, Scleroscope, Vickers scales.
Softness — reverse to hardness, is measured without causing plastic surface deformation.
Temperature coefficient of ultrasonic velocity — change of HF acoustic velocity with temperature; expressed for type of wave such as longitudinal excitation.
Thermal growth — increase in size arising from thermal cycling.

Pertaining to other needs

Zero shift — change of the zero state of the system, might be due to plastic deformation occurring in cyclic stressing, also regarded as caused by stress release processes as material ages.

Mechanical hysteresis — strain (real or apparent) retained during stress cycling, expressed as ratio of dimensions of the hysteresis loop; caused by many reasons including internal material properties, design of mechanism, internal heating effects. (Trylinski, 1971, Wiggin, 1972; Huntington, 1975; CDA, 1953).

Creep — slow dimensional change occurring, usually, under conditions of stress; most data relevant only for high stress, high-temperature conditions, not ambient, low stress situations found in instruments. One of the hardest parameters on which to obtain quantitative data for instrument design.

Internal friction, mechanical losses, damping — related to material viscosity, the loss mechanism within material structure; relevant to vibratory conditions; often temperature dependent. (Strong, 1938; Brush Clevite, 1966).

Rolling friction — resistance to rolling of one component on another; relevant to knife edge bearings and other precise rolling mechanisms. (Geary, 1955; Trylinski, 1971).

Coefficient of sliding friction — ratio of sliding frictional force to force normal to surfaces; always less than unity.

Ageing properties — variation of modulus of elasticity with time of either increasing or decreasing sign; due to release of internal stresses; term also used more generally. (CDA, 1953; Trylinski, 1971; Brush Clevite, 1966).

Set in coil, etc. — curvature existing in strip, wire after basic manufacturing process, needs to be accounted for in elastic element devices such as pivots; magnitude rarely quoted in data sheets.

Specific weight, density — weight per volume and per unit volume, respectively. (See Group F for other parameters that might also be placed in this group).

B. Electrical

Resistivity, specific resistance — property of material that impedes flow of current; expressed in two forms, linear and volume.

Conductivity — reciprocal of resistivity; should be qualified with word electrical to prevent confusion with thermal conductivity.

Thermal (or temperature) coefficient of resistivity, temperature coefficient of volume resistivity — variation of resistance with changing temperature, two forms of expression used.

Thermal emf, thermoelectric effect — voltage generated by the existence of a thermocouple; magnitude depends upon couple materials and temperature difference; significant where signals below 10 μV are important.

Pyroelectric effect — related to the piezoelectric effect in which electric charge is generated in certain materials by the presence of heat; its existence is doubtful and may be really piezoelectricity. (Glazebook, 1922, Vol 2).

Piezoelectric constants — electromechanical constants coupling mechanical action with electrical in the reversible piezoelectrical process; several exist, *k* for coupling, *d* for strain, *g* for stress. (Brush Clevite, 1966; Klaassen, 1978).

Curie point of piezoelectric coefficients — characteristic temperature above which the property is permanently lost; decides the absolute maximum temperature of operation.

Dielectric strength — ability of material to resist electrical breakdown whereupon electrical insulating properties are lost.

Dielectric constant — charge density per unit electric field; in piezoelectric materials it depends on degree of mechanical constraint.

Dielectric loss angle — a constant, as tan δ, from which energy loss in capacitive assembly can be calculated; can vary with material conditions.

Electrode potential — voltage developed on the material when formed into an electrochemical couple with a standard reference electrode material such as hydrogen; relevant in corrosion and stray magnetic and electric-field generation studies.

Non-sparking material — that which does not produce electric sparks when in contact with any other substance; relevant to use of material in hazardous, flammable situations.

Semiconducting features — not all materials obey Ohm's law for electric currents; used for semiconductor devices; may be important in precise, small-signal electric devices where strict mathematical relationships are required.

C. Optical

Reflectivity, reflectance — accounts for energy reflected at the interface of two materials of different refractive index, depends on materials, angle of incidence, wavelength, surface magnetic field, etc.

Transmittivity, transmittance, absorption coefficient, decadic coefficient of absorption — relates to radiation energy lost due to passage through material; depends on material, thickness, wavelength, polarization, electro-optic properties.

Refractive index — ratio of velocity of radiation in material against that in free space; decides bending action at interface of two materials; may be temperature dependent.

Spectral normal emissivity — ratio of emission of energy from a surface to that which would be obtained for a 'black body' at same temperature; constant only for a small temperature range, varies with material and wavelength.

Plane of polarization — plane containing incident and reflected beams, this being perpendicular to plane of vibration.

Seed grade, granularity, Schlieren quality, birefringence, striae, stones, crystalites — various terms describing manufacturing defects causing imhomogenities in optical material.

Abbe's number, v — value, reciprocal dispersion, constrigence — terms and constants describing how refractive index varies with wavelength in a given material; definitions vary.

Glass code — number assigned by industry to identify optical glass type.

Ratio of intensity of fluorescence — comparative ratio of intensity of fluorescence to that of a stated reference material.

Photoelastic stress constant — expresses usefulness, or otherwise, of material in photoelastic situations.

Verdet constant — angle of rotation of linear polarized ray in an isotopic transparent material passing through longitudinal magnetic field.

Resistance to radiations — damage statements made for various dosage and how it effects given properties; for high energy protons, low energy protons, electrons, neutrons, ultraviolet, weak X-ray, etc.; effect depends on temperature.

D. Thermal (Temperature coefficients are listed in the regime of relevance.)

Thermal conductivity —ability to transfer heat energy through a material by conduction; can be expressed as a coefficient for a given material.

Thermal diffusivity — coefficient combining thermal conductivity, density and specific heat values of a material.

Specific heat — amount of heat energy needed to raise the temperature of a unit mass of material by a unit rise in temperature, expressed in several forms; usually compared with water.

Melting point — temperature at which material changes substantially from solid to liquid phase.

Softening point — temperature at which the mechanical properties of material begin to alter; some materials do not melt but pass through this point to begin transformation into the gaseous phase, e.g., carbon, silica.

Film coefficient of heat transfer — constant expressing heat transfer across the interface between two materials usually of different phase; often the most significant barrier to heat transfer in a system.

Service temperature — that at which the material can be used continuously; may be expressed as upper and lower limits.

E. Magnitude

Permeability —ratio of magnetic field induced in a material to that magnetizing it.
Initial permeability — that existing in region of zero magnetizing force; the slope of the B–H curve at the origin; depends upon alloy and material heat treatment.
Saturation induction — maximum flux density that can be induced into a material.
Coercive force — opposing magnetizing force needed to remove previously present magnetization; depends on composition and heat treatment.
Magnetostriction effect — change in volume of material in presence of magnetic field; only contraction (or only expansion for some alloys) occurs for either polarity of longitudinal magnetization; is temperature dependent.
Curie point — temperature above which a material loses magnetic properties.
Magnetic hysteresis — expression of energy loss for magnetic material undergoing cyclic magnetic field reversals; usually given as graph of hysteresis loop.
Residual, remanent magnetism or remanence — magnetic force retained by a material; falls off with rising temperature.
Hall's coefficient — expresses degree of voltage generated between sides of a material when electric current flows in it where magnetic field is present.

F. Miscellaneous (Further factors that may be relevant).

Availability for use because of cost, supply factors.
Does documented history of the batch to be used exist?
Have the properties of interest been qualified?
Is the material machinable; do any processing restrictions exist?
What is the durability?
What is the cost? Special materials often have quite large minimum order values.
Is the corrosion resistance adequate?
What are the cohesive and adhesive forces for the material?
Chemical properties may be important.
What are the heat treatments needs, if any?
Are suitable stock forms available?
Is moisture absorption important?
Can the materials be soldered or welded where necessary?
Are suitable adhesives available?
Does the material have satisfactory aesthetic appeal?

To what extent are the various properties reversible?
Will it wear well?
How toxic is it?
How pure is the so-called pure metal?

Some special-purpose materials

Knowledge is assumed of the well used properties of the commonplace materials such as steels, aluminum and copper derivatives. They are adequately covered in mechanical and other relevant engineering design texts.

All of these play an important part in instrument system design where they are used in the conventional manner to provide gears, slides, bearings, frames, cams, shafts, chains and the like. A brief description follows of some special-purpose materials that find application almost solely in instrument construction, and therefore tend not to be described in general texts.

Low thermal expansion materials — Many assemblies require tight control of dimension while operating in varying temperature conditions that will alter size and shape due to the thermal coefficient of linear expansion of the structural materials used. (The various strategies available to overcome this problem are discussed in Chapter 7.)

The normal constructional materials, such as the various steels (coefficient of thermal expansion varies around 11 to $16 \times 10^{-6}/°C$ for different types), aluminum (coefficient around $25 \times 10^{-6}/°C$), and the brasses and bronzes (coefficients around $18 \times 10^{-6}/°C$) are too thermally responsive for some applications. Three dominant groups of materials can be used to exploit their lower value coefficients. It is not, however, always desirable to use the lowest possible thermal expansion materials; for example, metal to glass seals require from 5 to $10 \times 10^{-6}/°C$ coefficients. The other quite different properties of these three groups make them suitable for different applications.

In the first group are those based upon the alloy of nickel and iron plus quite small additions (or unwanted presence) of manganese, silicon, copper, chromium and cobalt. The alloy mix can be tailored to provide a wide range of desirable special properties including low expansion and, as will be discussed below, low variation of the elastic modulus with temperature change, with both properties optimized simultaneously.

Low thermal expansion occurs for the ratio of percentages 36:64 for Ni:Fe (the additives being included in the iron part). The actual coefficient obtained is a function of the alloy mix and the material treatment. It is not constant with absolute temperature, usually being made to exhibit the lowest value around room temperature. Cold working further reduces the annealed state coefficient from around 2.0 down to $0.8 \times 10^{-6}/°C$, but this will then gradually decrease again as the stresses are relieved over periods of years. Artificial ageing, a costly process, can be used to stabilize the coefficient. It can even be made slightly negative in sign. This material is marketed under such trade names as INVAR 36, (Telcon, Hamilton Technology) and NILO 36 (Wiggin); no doubt some other suppliers exist but there will not be many more. The lowest value material is sometimes called SUPERINVAR. The low expansion with temperature advantage is not gained without loss of other features; the elastic constant is less than that for steel, and it is hard to machine due to pronounced work-hardening. No other basic metal alloy or metal element is known that exhibits such low expansion.

The next group of materials that might be usable to gain low expansion characteristics are those based upon silica.

Normal glasses have coefficients ranging from 3 to $10 \times 10^{-6}/°C$. The borosilicate glass coefficient is nominally $3.5 \times 10^{-6}/°C$. Fused silica, often synonomously called (in slight error) fused quartz, has a coefficient in the range 0.22 to $0.56 \times 10^{-6}/°C$. Note the additional places of definition which are possible due to the constancy of the coefficient, the greater amount of research that has been expended on this material, and the possibility of using optical methods of dimensional measurement within the material itself to measure the expansion coefficient.

Where fused silicas are still unable to supply low enough expansion characteristics, designers can use other silica products. ZERODUR, by Schott of West Germany, is a glass-ceramic which is comprised nominally of 70% of randomly distributed crystals of high-quartz type in a residual glass phase, the two together exhibiting 'practically without thermal expansion' properties. The coefficient can be tailored to be as low as $0 \pm 0.05 \times 10^{-6}/°C$ over the range of 0 to 50° C. The values given are limited by the resolution of experimental method.

The alternative choice might be to use the ULE (ultra-low expansion) by Corning (Code 7971) which is a true single-phase glass, not a glass-ceramic. It is made from titanium silicate. Its expansion coefficient is given as $0 \pm 0.3 \times 10^{-6}/°C$ for the range 5–35° C; wider ranges are tied to values 0.03 ± 0.03, being negative below room temperature.

Both ZERODUR and ULE materials are made basically for optical work and are produced in optical blank shapes, (round discs, sheets, squares, rectangles). Small rods, and other instrument part shapes, are not so easy to procure. Fused silica, however, is sold in rods, tubes and fine canes, and it is generally easier to employ. Silica derivatives are brittle, but the apparent fragility of glass-like substances should not deter the designer from giving them consideration. In many instances they are quite tough enough, especially the 'space-age' glass ceramics. The working of silicas requires quite different machines and processes from those of the metals, but a well trained glass-blower can generally provide a suitable component, provided the design allows for these differences in processing.

The third group of low expansion materials that might be useful are those based on the use of carbon fibers. Carbon has a negative, small thermal coefficient of expansion ($-0.73 \times 10^{-6}/°C$ over 25 to 100° C), but this basic element in its macroform has very poor mechanical strength properties. In the 1960's research led to the commercial production of fine fibers of virtually pure carbon which exhibit an elastic modulus that is double that of steel while being very light, chemically inert, and having many more desirable features. The problem was how to make use of myriads of micrometer size filaments. Methods were devised to incorporate carbon fibers into a suitable matrix in which the bulk of the properties are retained and the fibers are made to act in unison as a macro material.

A main interest in carbon fiber methods has been in their use for making turbine blades, but the material has found its way into instrument applications on occasion. In Britain a very large throat size micrometer (around 2 m) was made using carbon fibers in a matrix to form the yoke. It was quite light for its size, more rigid than steel could be made for the same weight, and less influenced by thermal defects caused by handling and the ambient environment. The use of carbon-fibers has not been widely exploited in instrument design, and it seems that there is still considerable scope for their use.

The thermal expansion of various materials is shown in Figure 2.2. The graphs are only illustrative and should not be used for accurate design.

Constant modulus (of elasticity) materials — The nickel-iron alloys can also be mixed in different proportions to obtain special properties other than, but *not* coincident with, the lowest thermal expansion possible, an

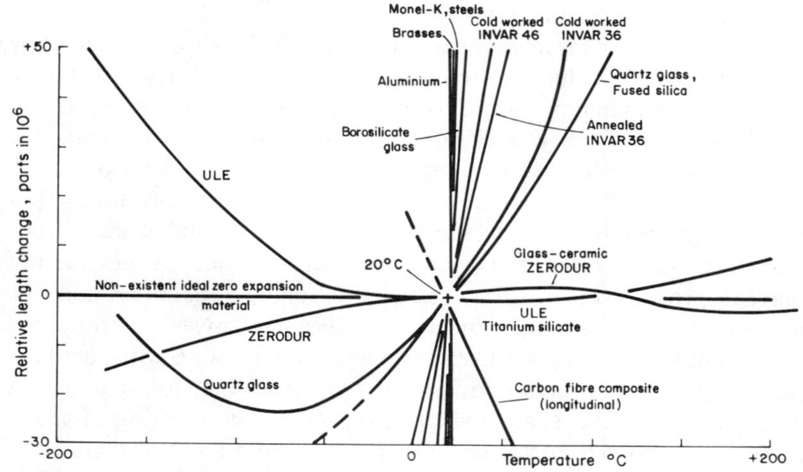

Figure 2.2 Thermal expansion, referenced to 20° C, of materials used in instrument design (Various sources combined).

important one being to have virtually zero change in Young's modulus of elasticity with temperature change. This valuable property (see Figure 2.3) occurs when the nickel content is around either 27 or 42%; the latter is generally used. Additions of small amounts of other elements enable this property to be tuned with regard to magnitude and temperature of operation and give the alloy better strength properties. Heat treatment is then used to modify the magnitude from virtually zero upwards — not all applications require a zero modulus of thermoelasticity. Post-melt treatment (Figure 2.4) also enables the mechanical hysteresis to be reduced signifi-

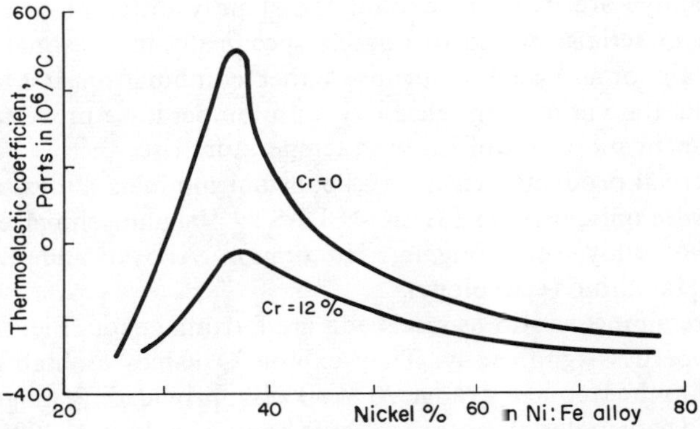

Figure 2.3 Thermoelastic coefficient of various nickel proportions of nickel-iron alloy with varied chromium content (Wiggin) (©INCO EUROPE Ltd. Used with permission.)

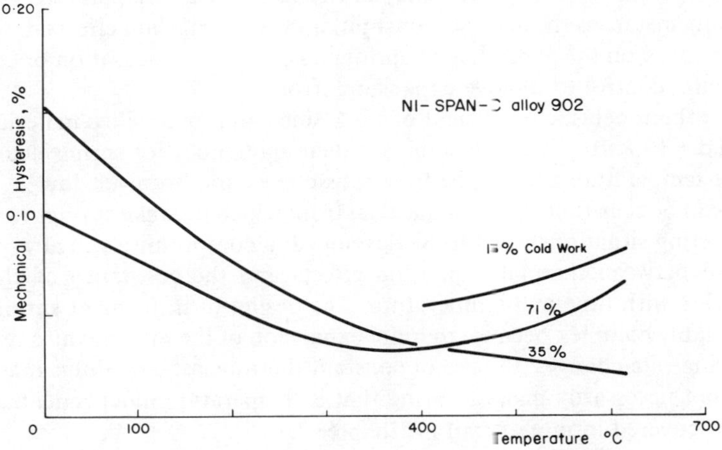

Figure 2.4 Mechanical hysteresis of NI-SPAN-C™ alloy 902 for 5 h heat treatment at temperature and degree of cold-working shown (©Huntington Alloys. Used with permission.)

cantly from a typical 2% for common spring alloys down to a low 0.03%. (Hysteresis percentage is the ratio of the maximum width of the load-deflection hysteresis loop in the vertical direction to the maximum height of the loop, expressed as a percentage.)

Normal steels vary in elasticity at $-250 \times 10^{-6}/°C$; copper spring alloys vary at the rate $-360 \times 10^{-6}/°C$. Constant modulus alloys can be made with only a part or two per million constants. This means that normal springs vary −1.0% for a 20−40° C change compared with springs made from these alloys which vary by only ± 0.01% for the same temperature interval.

Three alloys are made to exploit the slightly different constant that occurs for materials formed to operate specifically in torsional strain, in tension strain, or as a general purpose 'either' combination. It is also possible to make the variation in elasticity with temperature move to become either stiffer or more compliant with temperature rise.

Commercial products marketed as constant modulus alloys, usually as strip and wire only, include THERMELAST (Vacuumschmelze GmBH), NI-SPAN-C alloy 902 (Wiggin, Huntington Alloys) and ELINVAR EXTRA (Hamilton Technology).

They are almost as stiff as steels and are ferromagnetic. Field alters the elastic properties significantly. They exhibit Q values as high as 15 000 when used in the frequency range 20 to 50 kHz, provided no magnetic field is present. These materials possess a Curie point of only 120 – 190° C, above which the constant modulus feature is lost.

Temperature coefficient of linear expansion for these alloys is around 6 to $10 \times 10^{-6}/°C$. No alloy has apparently yet been devised that combines very low expansion with constancy of the elasticity as temperature varies. In many instances the designer must put up with expansion effects in order to capitalize on the constancy of springiness, using compensation or environmental control to remove expansion errors.

The thermoelastic coefficient of ULE silica and fused silica materials is around $+ 10 \times 10^{-6}/°C$, which makes them quite poor for spring elements where temperature varies. The hysteresis effects are, however, low.

It can be seen that the ideal material from which to make springs for use in exacting situations is yet to be developed; a compromise will always be needed between thermal expansion effects and the constancy of elastic behavior with time and temperature. The design of instrument springs is reasonably complex because thermal expansion of the material also results in spring-rate changes; the use of constant thermoelastic modulus material does not necessarily make a spring that is temperature independent. This will be covered in more detail in Chapter 7.

Diaphragm materials — Diaphragms, membranes, and capsules are used in measurement in pressure gauges and transducers and as elastic supports of various kinds. The materials may take the shape of thin diaphragms, corrugated multiple or single bellows, or may be used in Bourdon tubes. The majority of these elements are made from metals, but fused quartz has been used to form a Bourdon tube of a very sensitive pressure transducer.

Clearly, the elastic characteristics of the element are important but to them must be added other possible salient requirements such as high corrosion resistance, impermeability to the fluid held back on one side, high operating temperatures, high ductility for the forming operation, good weldability, and low mechanical hysteresis. These parameters mean that materials mentioned in the above two groups are often not suitable and a compromise must be made between the various shortcomings of a wider choice of suitable metals.

Diaphragms are made from brasses (the 80:20 ratio of copper:tin is generally used), stainless steels (such as AISI 316), constant modulus thermoelastic alloys (such as NI-SPAN-C), beryllium-copper, phosphor-bronze (a copper-tin alloy), high percentage nickels with significant chromium content (such as Hastelloy C), Monel (a copper-nickel alloy) and others. Trylinski (1971) and CDA (1953) give some general information on the materials and forms that are used; more detailed information can be obtained from metals and bellows makers.

Because these elements must be made to exacting measurement performance, as well as acting as enclosing elements, they demand well controlled dimensional tolerances in manufacture. It is this parameter that largely distinguishes diaphragm materials from others of similar composition.

Magnetic materials — Instrument applications for magnetic materials range from those needing highest permanent magnetic field strength to those in which the material must be magnetically very soft so that the part does not become magnetized. The range of need is very wide and there are many alloys to choose from.

Permanent magnets of high field strength are generally made from high nickel-cobalt alloys which, having melting points among the highest of alloyed metals, complicate production; they must usually be cast in their final shape and then ground to final size. Cheaper permanent magnets can be made from sintered ferrites, but these cannot provide the same level of performance.

Materials for making magnetic shields, such as those needed around cathode ray tubes, require high permeability (so that they 'shunt' stray fields) and high electrical resistivity so that eddy currents that might regenerate magnetic fields are minimized. Probably the best known is Mu-Metal (77Ni:5Cu:2Cr) which has comparatively very high electrical resistivity with a reasonably high permeability and low saturation induction. Presumably, it is not possible to obtain all desirable coefficients simultaneously; shield design can be quite taxing due to this and because of the varied situations that can occur (Arendt, 1966). At least 20 quite different magnetic alloys are available, each developed for a particular reason.

Magnetically soft materials, those in which magnetism is not required, are useful for laminations of magnetic circuits. In instrument transducers it is reasonably pure nickel that gives the best magnetostrictive properties, as shown in Figure 2.5. Good response also occurs with 20% and 4% nickel-iron alloys. The 45% (50% is actually used in practice) peak is often used to obtain better hysteresis and eddy-current losses at the expense of less sensitivity. Thermomagnetic alloys (70 Ni:30 Cu), Figure 2.6) are used to compensate for magnetism changes through their changing permeability with temperature, CDA (1953).

One other kind of material that might be placed in this group is that developed to have nonmagnetic properties along with high wear and strength features, for example, the gapping material in a tape recorder head. A typical material for this kind (such as HAVAR, by Hamilton) can

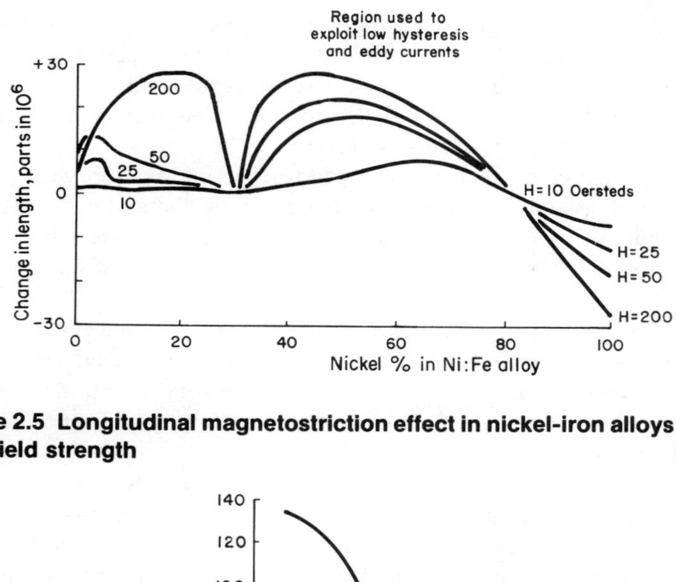

Figure 2.5 Longitudinal magnetostriction effect in nickel-iron alloys for varying field strength

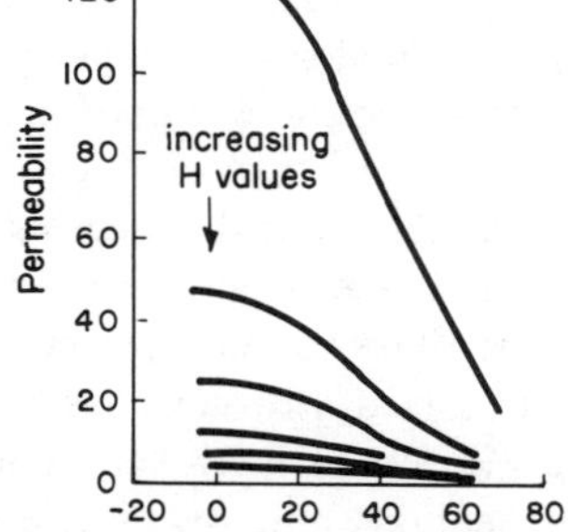

Figure 2.6 Change in magnetic permeability with temperature and field strength for J.A.E. metal (Wiggin) (©INCO EUROPE Ltd. Used with permission.)

exhibit a permeability of 1.00004 and retain virtually no detectable magnetism after high magnetic field exposure.

Once again, although the designer appears to have a large variety of materials to choose from, very few will be ideal and the choice soon narrows to one or two as all specifications are checked.

Electrical resistance alloys — Electrical resistance wires, tapes, and foils play a direct part in the electrical circuitry of instruments. Required characteristics might be high specific resistance to reduce the size of a resistor, constant resistance wih varying temperature when used as a shunt, or high operating temperature when used as a heating element. Another factor needed might be a known thermocouple voltage-to-temperature relationship.

In strain gages and precision resistors, modern methods of manufacture make use of ultra-thin foils and photomechanical etching procedures to form the gage and resistance elements. Foils for these applications are often supplied to resistance controlled tolerances rather than to dimension. Foils are chosen by the designer to suit the application and the substrate onto which they are to be attached.

At least 25 different common resistance alloys are available. Their temperature coefficient of resistance varies from $\pm 4500 \times 10^{-6}/°C$ down to $\pm 10 \times 10^{-6}/°C$. Thermal emf versus copper values range from -39 $\mu V/°C$ down to $-1\ \mu V/°C$. Resistivity similarily varies widely, spanning a variation of 40 times the minimum. This range may appear to give the designer any choice that is needed but, as with other groups already mentioned, these too occur in only certain combinations for which there is usually one or more less than ideal characteristics that must be tolerated.

Resistance alloys generally possess a thermal coefficient of expansion in the region of 11 to $18 \times 10^{-6}/°C$. They are formed primarily as alloys of Cu:Ni; Cu:Ni:Mn; Ni:Fe; Ni:Cr; Ni:Cr:Al; Ni:Cu:Fe; and Fe:Cr:Al.

When selecting materials, as with all other classes, it is important to establish what other parameters of an unusual nature might be relevant (using the check list given above in the previous section) and then seek out the appropriate constants and coefficients from data sheets and manufacturers.

State-of-the-art designs will often need experiments on selected materials to establish these parameters for they may not have been of any relevance previously and are therefore not determined.

References

Arendt, C. H. Jr. 1966. *The when, why and how of magnetic shielding*, File No 52000/B9236/F6, Westinghouse Electric Corp, Blairsville, USA.

Braddick, H. J. J. 1954. *The physics of experimental method*, Chapman and Hall, London, UK.

Brush Clevite. 1966. *Piezoelectricity — a data book for designers*, Brush Clevite, Southampton, UK.

CDA. 1953. *Copper in instrumentation*, Pub No 48, Copper Development Association, Radlett, UK.

Drysdale, C. V. and Jolley, A. C. 1924. *Electrical measuring instruments*, 2 parts. Ernest Benn, London, UK.

Faires, V. M. 1955. *Design of machine elements*, Macmillan, New York, USA.

Geary, P. J. 1955. *Knife-edge bearings — a bibliography survey*, (Survey of instrument parts — No 2), Sira, Chislehurst, UK.

Glazebrook, R. 1922. *A dictionary of applied physics*, 5 vols, Macmillan, London, UK.

Hamilton. 1975. *Metals — development, processing*, Hamilton Technology, Inc, Princetown, USA

Huntington. 1975. *NI-SPAN-C alloy 902*, Pub No 2M 8-75 T-31. Huntington Alloys, Huntington, USA.

Klaassen, K. B. 1978. 'Piezoelectric accelerometers', in *Modern electronic measuring systems*, ed. P. P. L. Regtien, Delft University Press, Delft, The Netherlands.

Strong, J. 1938. *Modern physical laboratory practice*, Blackie and Son, Glasgow, UK.

Titcomb, G. R. A. 1966. *Fundamentals of engineering science*, Hutchinson Technical Education, London, UK.

Trylinski, W. 1971. *Fine mechanisms and precision instruments — principles of design*, Pergamon Press, Oxford, UK, translated by A. Voellnagel from Polish original.

Vacuumschmelze. 1977. *THERMELAST 4002, 5409, 4290*, Pub. No. F001-3/77, Vacuumschmelze GMBH, Hanau, FR Germany.

Wiggin. 1972. *Wiggin nickel alloys for electronics and instrumentation*, Pub. No. 3571A, Henry Wiggin and Co. Ltd, Hereford, UK.

Chapter 3
Fundamental concepts of systems nature

The purpose of mechanical design

Mechanics is the subject that deals with the study and application of the motion and equilibrium states of material bodies. The aim in mechanical instrument design is to create a specific mechanical system that provides the necessary controlled movements in such a way that predetermined moving and fixed relationships are preserved under reasonable conditions of operation.

Reuleaux, in his *Kinematics of machines* of 1875, was apparently the first person to see machine design as more than a complete design that was modified until it worked. He defined a machine as "a combination of resistant bodies so arranged that by their means the mechanical forces of nature can be compelled to do work accompanied by certain determinant motions." He further expanded that motion of parts is simply determined by the geometry of the system of parts (Burstall, 1970). Reuleaux's generalizing definition applies equally well to mechanical instruments which are a form of machine.

Departure from the ideal concept will occur for many reasons, for example elasticity of component members, clearances and wear in joints, and 'second-order' influences such as thermal expansion of the material. Design is the process whereby a satisfactory compromise is made between the range of envisioned parameters to come to a single arrangement. The designer strives to produce a design statement that will lead to an article being made that performs the set task within the stated specifications of performance without the need to remake the article after it is first assembled.

Mechanical design has two main stages of progression. The first is where the kind of necessary mechanisms are realized as a system. The following stage then selects a method of creating the chosen individual mechanisms as hardware that will suit the total task. For example, the mechanism seen to be needed may be a means of rotating one shaft at a different speed to another while maintaining a strict gearing ratio. That is the first step. Then follows study of the various ways of achieving this — toothed belts, friction gearing, spur gears, and others — to see which suits the case in hand.

No matter which type of mechanism is involved the fundamental issues behind the choices of the first step are concerned with controlling the freedom of movement of selected parts and points. This will be expanded later.

The design of the instrument, that machine used for handling informational rather than energy quantities, generally requires specific attention to the fidelity of the input-output and other relationships over extended periods of operational life. There is less interest in efficiency of energy transfer (although in mechanical design, losses due to friction can cause thermal changes to components that give rise to error). Instrument design is generally dominated by attention to deflections of the system and its components rather than to their strengths.

A general introduction to mechanical machine design can be obtained from the widespread works describing the design of machines that can transfer energy, examples being Faires (1955), Beggs (1955) and Lent (1970). These accounts will, however, only take the user a limited distance into instrument design, for they generally lack specific detail about fine mechanism requirements.

Works which discuss the general systems level of design of instruments are few. They include Glazebrook (1922) Vol. III, Rolt (1929) Vol. II, Whitehead (1934), Strong (1938) and Braddick (1954). This chapter is provided to help overcome the difficulties in procuring information about the modern approaches to design that are now possible.

Factors important to design include the methodology for controlling the design work itself, the cost of producing the design, building the instrument and servicing it, the degree of serviceability; the ease of measurement in production; the needs for calibration after manufacture; and ergonomics requirements (now often called the man-machine system — MMS). These and others must all be properly thought out so that the users get correct results. Many of these vitally important factors are often neglected in the initial design, being added by adaptation or modification at a later stage when the design has progressed so far into production that major change cannot be entertained. For example, a firm making fever thermometers only realized that their product must fulfil certain calibration criteria when they were ready to launch the product onto the market place!

At the present time, instrument design is largely a case of some intelligent application of science plus a very large component of experience through years of actual involvement with similar products. The author's views are expanded elsewhere (Sydenham, 1979).

Design procedures for each of the different physical regimes of an instrument have their own peculiarities. Mechanical design must allow for the fact that a wrong choice is not as easily corrected as in, say, electronics. There is, for example, no simple mechanical counterpart of the variable resistor of electronics, which can be placed into the system, to find an optimum value, with great ease and with little cost or performance penalty. To obtain adequate performance from a mechanism usually requires very close tolerances. Dimensional parameters in mechanical systems will often require parts per million precision compared with parts per hundred in much of electronic technique.This means that the various parts of a mechanical system must be made precisely, or at least be capable of being assembled to close tolerances, if adequate performance is to be achieved. Control of wear is important since it may degrade performance.

Electronic design, in contrast, can more easily compensate for changes to the value of components, since they have been initially burned in to eliminate the few that will fail.

On the credit side, mechanical design is more intuitive and readily understood; wrong factors in a mechanical system often show up very easily without the use of sensory aids because mechanical systems interface more directly to the human sensors than concepts of other disciplines. Mechanical resonant systems can also yield very high-Q oscillation for low component cost.

An important feature of mechanical systens, which may be either advantageous or deleterious, is the inertia of components. The necessary inherent mass of mechanical parts is very often the limiting parameter to dynamic performance when high frequency operation is needed. It can, of course, be a useful factor providing integration properties. In a multiregime instrument, the mechanical parts will often dictate the upper frequency limit. Generally, unless specifically and carefully designed to be otherwise, mechanical systems will be low-pass in performance.

Specifications

There exists a maxim "$pecify $anely or $uffer". It sums up the importance of spending enough time on raising adequate and realistic specifications on which to base the design. Mechanical design does not differ from any other design in that to get started one must have a set of specifying statements that will outline, in a quite rigorous and meaningful manner, what the instrument must do. These statements are called specifications. Too often, instruments just happen with these growing as the design proceeds. A proper set of 'specs' at the beginning of the design will aid the designer to realize what is ultimately required. They should be set to paper *before* in-depth design begins, for once the designer has become immersed in the design it is a fact of life that he or she can become very resistant to change.

Specification should not be raised by the designer or user alone but as a group effort. Perhaps there should be a joint agreement between three parties — the system end-user, the system designer and manufacturer (if they are of the same group), and the subsystems and component vendors who supply the parts that will be used, generally without modification. Figure 3.1 shows how Wheeldon (1974) depicted this process. His two papers expand the following points now given. Various kinds of specifications exist, each having a different purpose of definition.

Systems user and systems designer relationships

- Operational requirements specifications
- Functional and technical requirements specifications
- Design specifications (to implement and check design)
- Factory test specifications
- Acceptance test specifications (Factory)
- Documentation specifications
- Packaging and delivery specifications
- Installation specifications
- Commissioning specifications
- Acceptance test specifications (Field)

Systems designer and component vendor relationships

- Standard item specifications
- Modified 'Standard' items
- Special subcontract specifications
- Incoming inspection specifications (System designer)
- Incorporation of vendor specifications in system design specifications

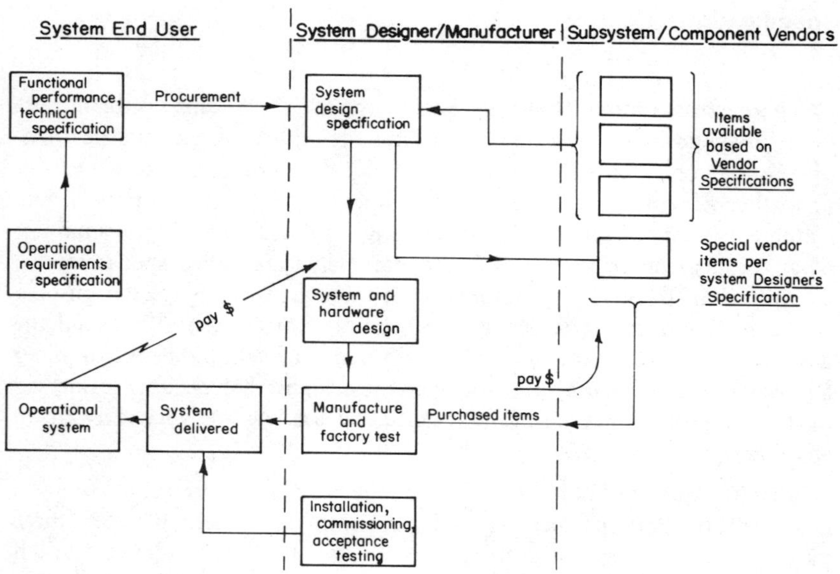

Figure 3.1 Vested interests in a typical system design process (Wheeldon, 1974) (Courtesty BWD Electronics Pty Ltd)

The designer is the person who has to provide specific detail to the factors presented by the various groups. A realistic producible design at a satisfactory cost must be achieved. Value judgments are needed about what the customer says is wanted (which is often what is then *thought* to be needed); what the vendor can offer (or what is *said* to be available); the designer's own ability and facilities for design; and the allowed time and money available to complete the task.

Unfortunately, to specify a product completely before it is made requires powers of prediction plus enormous effort in attempting this virtual impossibility. To attempt to create a total specification that gives the user complete assurance of being absolutely safe from faults and failures of compliance is an unreasonable aim. The cost and time needed are prohibitive. A compromise is therefore struck that specifies the product in a short form manner in reasonably standard 'shorthand' statements.

Specifications, therefore, attempt to portray what both the designer-maker and the purchaser-user expect to produce and receive, respectively. This verbal interface is poorly developed for the smaller systems and the stand-alone instrument. In the larger extensive instrument systems, the complexity there makes better assessment necessary, but then the complexity prevents absolutely adequate statements from being realized.

In practice, dissatisfaction often arises between the parties to the contract and, in large financial value cases, independent arbiters are called in to make judgments. The art and science of contract and specification writing are now extensive; specialists exist to cover the needs. It is beyond the scope of this book to go deeper but here are some simple rules to follow.

Try to use specifying statements that are the result of past experience. Examples are Codes of Practice and the various Standards specifications. For instance, reference to the BS code for dimensional tolerances will soon show if those called for are reasonable for the size of the part. The procurement of extensive systems of measurement equipment is covered in CAMA (1974) and sections of Antill (1974), which, although not on mechanical apparatus, specifically deals with matters of contractual statements.

Always watch for 'minor' details in a statement that may be more major than they are thought to be. For example, surface finish to be within a given limit may be specified for a pressure cell capsule, yet nothing is given about the fluid tightness of the material. Further, never leave important factors out because they seem so obvious. The converse holds; if they are not given, check them out.

All statements in a specification should be carefully studied *before* signing an agreement; it is of little value to look into it afterwards. Look carefully for fundamental inconsistencies. Does the frequency response called for exceed that of any mechanism yet produced? Does the required or specified noise level fall below the thermal control errors? Suppliers cannot make an instrument that refutes the known physical laws; but they have been known to offer such for sale — unwittingly, of course.

Make clear what specifications apply to what, at what stage. Care is needed to prevent them 'creeping' in the hands of enthusiastic clerks. There was a case of an optical shaft encoder that had to read to arc-minute accuracy when mounted on a radar tracking head. By the time the specifications were transferred through several stages of contract writing and then into the maker's plant they had crept to stated need for arc-seconds performance. The manufacturer, not wishing to be caught out, went better again. The result was that the encoder cost about a hundred times the price of one that would have done the job. Furthermore, the full encoder accuracy could not be used because the radar head was not designed for better then arc-minute use.

Specifications can also come to the designer as the result of the so-called 'horse-camel' syndrome. The end user collects data sheets of the various products offered by several different sources. From each of these he picks out the individual good points, adding these together to raise his own specification list. This process entirely ignores the fact that these special features are often the result of the designer trading off one feature to allow another to be enhanced. The result of this combinational method is a specification of an instrument impossible to make. Unfortunately there are many suppliers in the instrument manufacturing industry who lack enough experience to recognize that such tenders may be impossible to make and they take them on only to find later (to their and the purchaser's regret) that they had been attempting a virtually impossible task. The use of independent expert consultant groups, a well trained in-house team, study contracts, and part-stage costing may help.

At all times it is vital to 'clear the air' between all groups involved. If the user believes the designer cannot do what is suggested, then it should be challenged very early. Similarly, the designer should state what is unclear early in the negotiations with the consumer. Change becomes progressively difficult as the designer is gradually forced to freeze the parameters in order to make concrete progress.

When writing specifications, considerable care must be given to the provision of statements that are meaningful and not ambiguous. Unfortunately, to write a reasonable specification requires the use of short-form statements. For example, statements of accuracy or linearity are used very loosely. Graphs could be used much more to define such parameters as frequency response, linearity, and the like. When in doubt, seek further detail about the meaning of a parameter.

There is no harm in verifying the self-quoted performance of subcontractors and vendors.

All too often the user does not adequately think out the task for the instrument that is called for. The designer should be expected to ask poignant questions of the user, for the designer will often know the design problems more closely than the user. Very often discussions show that the user really wanted something quite different from what was thought to be necessary. Take, for instance, the person who once asked for a thermometer to measure length changes of a component because he did not see that it could be measured directly as a dimension.

Each case stands on its own merits. In measuring instrument design, the spread of applications is so broad and design becoming so complex that poor performance often results. The current state of instrument design and manufacture is such that it is suggested that instrument users must constantly be on their guard against the supply of faulty instruments — see Moss (1978) for a relevant report.

Instruments, especially one-offs and those with mechanical elements, are still in a state of low inherent quality capability compared with the performance that one now expects and usually gets when using electronic systems based on standard integrated circuits. If an IC works, it is going to be in specification and continue to operate. It will also be a very well designed component. In contrast, mechanical systems may continue to operate but with deteriorating fidelity.

Controlling the degrees of freedom

Mechanical design is fundamentally concerned with providing a system of materials arranged to allow controlled movement with from none to all six degrees of freedom of an extended object within a defining space.

A theoretical point in space has three degrees of freedom — the three translations that are generally expressed as x ,y, z in the cartesian system. Other coordinate methods can be used. The concept of rotation has no meaning for a point.

A point forming a position on an extended object, however, can have six degrees of freedom for, in addition to the three possible translations, there can be three axes of rotation about the point.

The essence of good mechanical design is to devise a system that restrains as many of the six freedoms as are needed while allowing freedoms. For example, the slide of a travelling microscope must allow one translation, and while doing so, constrain the other two translations and three rotations as perfectly as are needed. Lack of straightness of the slide will cause small translations and rotations to occur as it slides. The degree of precision needed dictates the necessary tightness in the tolerances of manufacture and the method by which this is accomplished in the face of wear, temperature effects, and other error sources.

Some other examples are the framework and sliding axes of a 3-D measuring machine (three translations allowed with rotations constrained) and the surveyor's theodolite (constrains all translations allowing the rotations).

An equally important case is that where all degrees of freedom are to be constrained. This occurs when mounting a component to stay in a chosen position relative to another; for example, a method of locating a survey leveling unit that is to be used repeatedly around a series of test lines.

At times, measuring machines are sometimes specified as, say, 3½ axes, meaning that they have a rotation allowed within the movement provided along one translation axis. It does not mean that the rule of six degrees of freedom for an extended point has been broken.

Two extremes of design philosophy exist for realization of a suitable mechanical structural relationship. Each has its advantages and disadvantages and often the design makes do with a compromise between the two extremes. These extremes are called kinematic and elastic designs; in between them is the semikinematic.

Kinematic design

When all parts of a structure are held in such a way that they retain positional relationships without being in a state of variable stress except at their points of contact, the design is said to be kinematic. In other words, effects such as a change in size due to temperature or material creep will not stress the frame. Clerk-Maxwell in the *Handbook of the Special Loan Collection of Scientific Apparatus* 1878 wrote this:

> When an instrument is intended to stand in a definite position on a fixed base it must have six bearings, so arranged that if one of the bearings were removed the direction in which the corresponding point of the instrument would be left free to move by the other bearings must be as nearly as possible normal to the tangent plane of the bearing.

Strong (1938) defines the principle of kinematic design as being applied when a body has at least 6-n points in contact with a second reference body when it is to have n degrees of freedom.

A spherical ball sitting in a triangular hole, as shown in Figure 3.2a, rests on three points of contact while it is under the force of gravity (which acts as the 'locator'.) Thus it has three degrees of freedom constrained with respect to the datum block (the translations) yet is free to rotate about three mutually perpendicular axes.

The classic example, demonstrating constraint of all six freedoms, is the so-called 'hole-slot-plane' or Kelvin-coupling arrangement. This is easily understood by reference to Figure 3.2b. The hole constrains three translations (three contact points), the slot constrains two rotations (two contact points), and the plane constraints the other rotation (one contact point). Addition of any extra contact points, unless made perfectly enough (as in a ball-race bearing), would add a 'redundancy' causing elastic deformation of the frame due to 'over constraint'.

It is easy to see that changes to the size of the top plate will cause the legs to move yet retaining the same stress relationships (provided the slotted plane contacts move accordingly). Note that the hole-slot-plane does not, however, retain the center of the top member coincident with that of the lower member if a relative dimensional change should occur betwen the two members. A variation of the 'hole-slot-plane' that uses three radiating slots (Figure 3.2c) overcomes the center shift defect. Figure 3.3a shows a kinematic form in which one degree (translation) is allowed, this being proven because five contact points exist. Figure 3.3b shows how this form is modified with rolling bearings to retain point contact while reducing wear. It is typical of the slide of a travelling microscope.

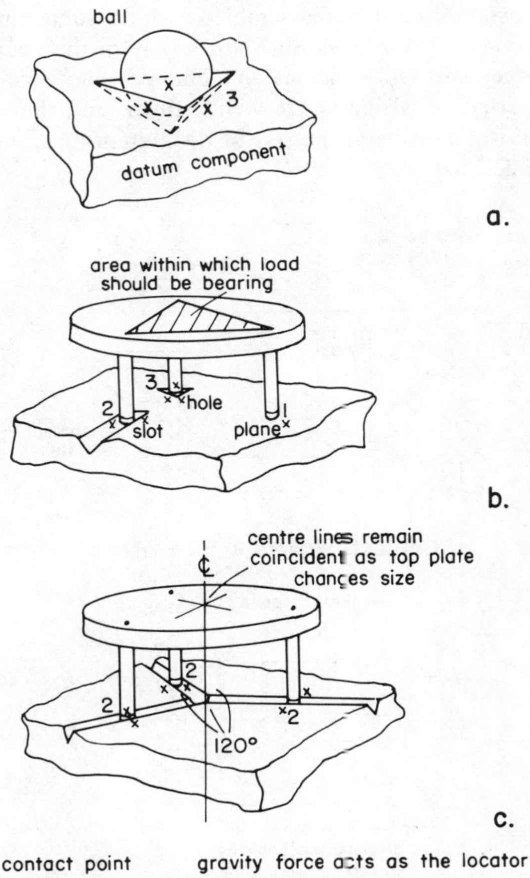

Figure 3.2 The degrees of freedom in a kinematic design of a coupling are equal to six less the number of contact points. (a) All three translations constrained. (b) Hole-slot plane in which all six freedoms are constrained. (c) Slot-slot-slot variation of case (b).

Many arrangements are possible depending on the need to allow or control a given number of degrees of freedom. Sketches of various couplings are given in Whitehead (1934), and Braddick (1954).

The principle of kinematic (or other alternative) design is applied whenever two assemblies need to be coupled. It is necessary, therefore, to break down the whole system into its basic couplings or mechanisms, tackling the design of each in a hierarchical order to ensure that over-constraint does not occur unwittingly at a later stage by the addition of a link in an incorrect way. To reflect this process, the discussion of mechanisms used is divided in the two following chapters on kinematic and on elastic elements.

Features of the kinematic method are that its parts can be made and installed or replaced without altering the positional system. Once made, the kinematic assembly will be reassembled with the same dimensional performance. Secondly, wear does not induce play in the coupling mode — it may, however, introduce dimensional change. Another feature is that the stresses exerted on all parts are well defined, and they can be designed virtually in isolation from the rest of the system once their contact loads have been decided.

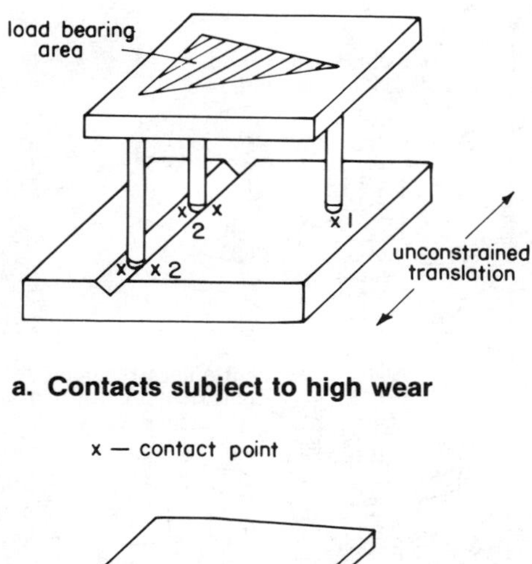

a. Contacts subject to high wear

x — contact point

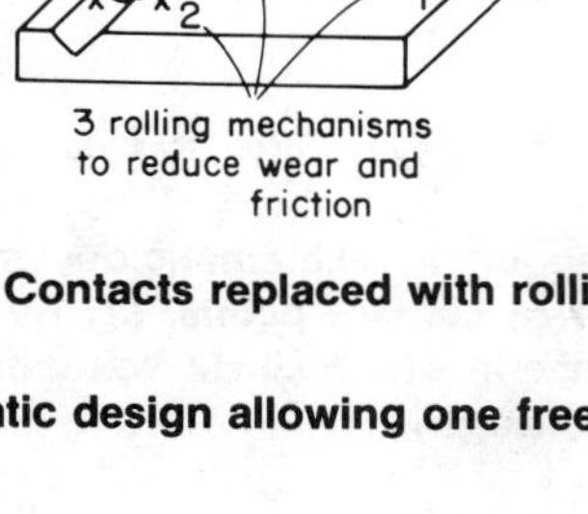

b. Contacts replaced with rolling devices

Figure 3.3 Kinematic design allowing one freedom of translation

Disadvantages are that there is need for some application of a locator force to hold parts together, such as gravity, magnetism, or small springs. The points of contact are also heavily loaded and consequently operate under conditions of high stress. The pure principle is often degraded by the use of extended contact surfaces to reduce the contact stresses — semi-kinematic design.

A further difficulty arises when it is seen that the systems result in triangular space-frames when redundant links are not present. The three-legged stool is a good example of the possible undesirable features of purely kinematic design. The stool always sits correctly on the reference surface without rocking, but place the load outside the central triangle defined by the leg positions and the linkage is soon lost. The locator should not be designed to withstand such forces.

Elastic design

The opposite philosophy is to join components rigidly, allowing them to flex within their elastic properties: a four-legged stool will still sit on any surface if the top is made flexible. The elastic philosophy deliberately contravenes the 'zero redundacy' law, doing so in a defined manner.

For this method to be designable the system must be so constructed that the designer knows just what flexes and where. In cases where only limited translation and rotation is desired, the elastic method is probably the best. Purely elastic design is almost an academic abstraction. In practice, the method becomes somewhere in between totally kinematic and elastic, being called 'semikinematic'.

As an example of the two extremes, consider the need to swing a pendulum or the beam of a weighing balance. The pivot may be made using knife edges that rock on a rigid flat (kinematic) or it may rotate on a single or crossed flexure strip pivot (elastic). Another example might be the need to support a sensitive tilt meter on a given surface in such a way that relative length changes, due to temperature effects on the body of the instrument, do not strain the instrument body. This can be done with either the kinematic hole-slot-plane concept or by the use of three elastic flexible legs that bend a little as the body changes length (see Figure 3.4). It is important to make specific calculations of the degree of flexing that is needed. What may appear to be a stiff rigid member can be comparitively flexible when the forces are high or the movements small. The designer should regard all components at the onset as being compliant, altering this belief only after calculation of strains due to disturbances shows them to be adequately stiff or compliant as the case dictates.

The mechanisms of the two groups of design — kinematic and elastic — will be dealt with in more detail in later chapters.

It is surprising that very few texts on general machine design actually discuss these basic design differences. It is in some of the works that are specifically concerned with instrument design that these concepts are expanded. Greater depth and more examples are to be found in Glazebrook (1922) Vol III, Whitehead (1934), and Braddick (1954). The choice of which philosophy to use and when is very much a matter of training and experience. Where elastic methods can be used they usually offer considerable cost and assembly advantages. For example, locating one part relative to another with, say, one allowed translation of short stroke can easily be provided with a parallel-strip suspension. It will be virtually free of frictional effects. It will be cheaper to make but does suffer in that a changing deflection force is generated as the 'slide' moves.

Wear-true philosophy

In 1913, in the *Aeronautical Journal* of July, Horace Darwin, co-founder of the Cambridge Instrument Co., expounded a concept of design that as mechanical elements wear through use they retain the dimensional precision needed — the wear-true philosophy.

The use of kinematic principles goes a long way toward this end, and by appropriate choice of proportion and arrangements, it is often possible to achieve a linkage that does not degrade accuracy as it wears or deforms.

Darwin illustrated this using the examples of the bearing arrangement of a wheelbarrow wheel. When the axle is fixed in the wheel, the axle ends wear round and the bearing hole wears oval, yet still round enough to give good location. But fix the axle in the shafts and the shaft wears oval in the wheel and the wheel bearings wear very large and sloppy.

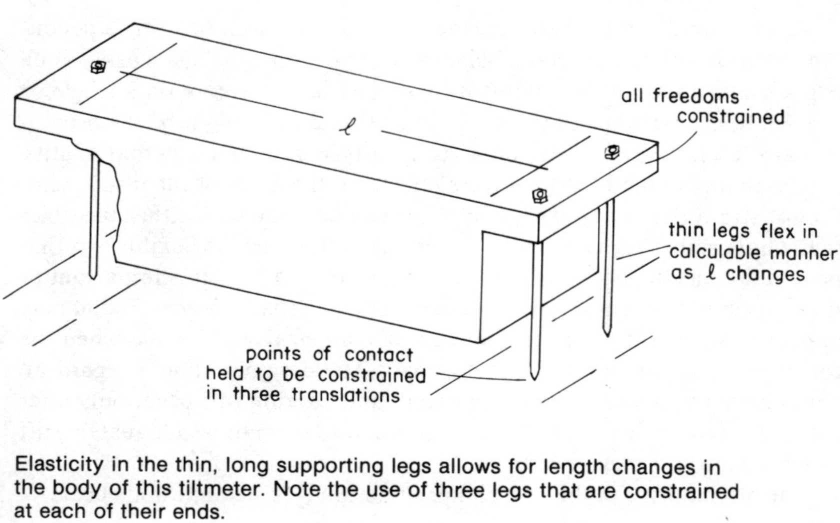

Elasticity in the thin, long supporting legs allows for length changes in the body of this tiltmeter. Note the use of three legs that are constrained at each of their ends.

Figure 3.4 Semikinematic design — (ANAC)

There seems to be no basic statable rule for achieving this and, although a general situation that occurs often is when the designer attempts to arrange wearing surfaces to wear evenly on each contact face of the sliding surface. Another approach is to make the wear occur at a geometric position that has the least effect in the geometrical precision. Figure 3.5 shows two designs, one of which does not adhere to the 'wear-true' philosophy.

Balancing of rotating parts is also another design requirement that can be placed within the wear-true philosophy; proper balance reduces bearing wear.

Darwin incorporated this principle in the design of his very successful 'rocking microtome' instrument of the 1885 period. It was capable of retaining ability to cut microscope specimen slices to micrometer thickness for a half century of service — see Cambridge Instrument Company (1945) and Braddick (1954).

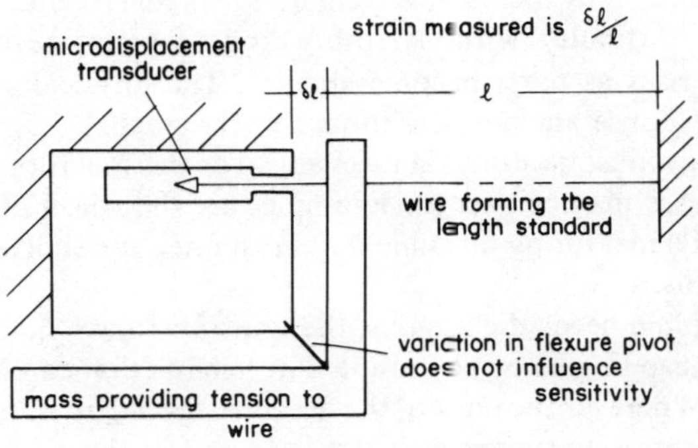

a. Wear-true philosophy adopted — pivot lengthening with time or abuse does not alter gain to displacement sensor

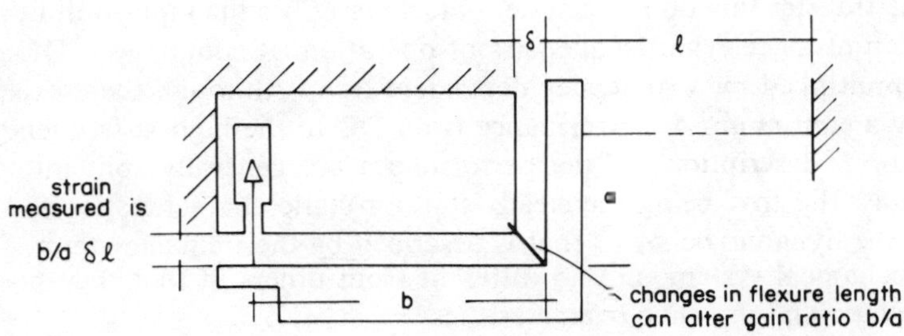

b. Dimensions of pivot directly influence mechanical gain

Figure 3.5 Two balances used to tension a wire length-standard in earth-strain meters

The size of an instrument

Unlike the situation occurring in electronic and in optical disciplines where size of components usually dictates the overall size of the instrument system, mechanical instruments may or may not have clear-cut reasons for having any particular size. So, at the outset of design, a choice must be made if progress is to take place that does not need to be completely revised later. Here are a few considerations that may help to decide the overall size of the instrument. Braddick (1954) provides specific case studies to illustrate some particular reasons for a given size.

Certain functions, such as a size dictated by a necessary component, may decide the basic size to be used. Examples are the objective size and focal length in a telescope, the length of translation of a measuring machine, the load and volume to be supported or contained, and the need to match a given existing product.

Portability may dictate low weight, but small size often means increased cost and difficulty with assembly and servicing. However, dynamic response rises as parts become smaller. The physical size of a high frequency response anemometer must be very small.

If energy must be stored in mechanical or thermal form, large rotating or static masses may be needed. Examples are the wheel of a gyroscope and the bed of a measuring machine that must integrate short-term temperature fluctuations.

Ergonomic needs dictate that the unit must have manual controls that suit man-size fingers or be mounted at heights that can be easily seen and reached. Whatever the reason, the designer should attempt to find objective bases for choosing a given size. If not, there will always exist the possibility of having to re-scale at a later date.

Statics and dynamics

The transfer function of a system block describes the input-output relationship at all relevant frequencies of operation. It comprises a DC gain term multiplied by a frequency dependent term. Although there exists in reality a continuum of performance from DC to the highest frequency of operation, descriptions of the performance are generally split into two domains, the first being the steady-state or static characteristics and the other the dynamic behavior that is described by the frequency term.

Mechanical systems are no different from others in that they possess both static and dynamic characteristics.

The design of a direct-reading travelling microscope requires virtually no understanding of the dynamics of movements that the operator imposes on its sliding parts unless it is large, in which case the inertia of the sliding part can be great enough to impart excessive forces restricting the operator's manipulation response. However, if the readout is automated by closing the translation drive loop, the dynamics of the whole system become important, if stable and rapid high-speed operation is to be obtained.

Apart from a very cursory statement saying that dynamics do occur in instruments, the majority of instrument design texts give little help with this aspect to the designer of modern multiregime instrumentation. Instead, the designer must read much more widely in order to become familiar with the basics of dynamical systems and then apply the general rules to instrument design where it is needed. (We return to this a little later).

Definition of static performance involves specifications being stated about such parameters as resolution of readout sensitivity to a given variable input, linearity, accuracy, precision, repeatability and many more concepts. These concepts, which are unfortunately not totally standardized in terminology, are described in most works on instrumentation and therefore need not be developed here. Doebelin (1975), Foxboro (1972), and Bell and Howell (1974) are examples. They are terms generally describing only the static behavior, the results of tests measured after the systems come to rest. During transients the performance can be very different indeed. In many cases the manufacturer provides no statement of dynamic behavior.

In the dynamic mode the designer must be aware of the effect of friction in its various forms, of the effects that the various kinds of slop between couplings produce: backlash, dead zones, and other nonlinear operation. Where the best acceleration is needed from a motor driving load, the load inertia must be matched to that of the driving source. A gear box is often needed to obtain this match. The transient performance of electric motors generally compared by the magnitude of their ratios of torque^2/inertia. This is called the 'power-rate' (West and Leonard, 1955, and Rothwell, 1966). The same term can be applied to any form of motor (spring motors, gravity feeds, and other mechanical motors) in a similar manner (Sydenham, 1969) giving the designer a chance to assess the relative merits of various mechanical or electrical drives. Where high acceleration is concerned, the diameters of rotating parts should be kept as small as possible, for rotational inertia increases rapidly with increase in diameter, sometimes as the fourth power of the radius. The problem of matching mechanical loads to sources is identical to the problems existing in electrical power transfer.

When resonance is to be invoked, or appears unwanted, the designer needs to know how to enhance or reduce it. The role of damping must be understood. In some instances the designer attempts to design for the highest Q conditions so that the system constantly 'rings' by noise excitation (a relevant example being in one form of gravity wave detector that uses large freely supported blocks of aluminum as detecting resonators). This provides increased sensitivity to signals at the resonant frequency. The dimensional changes, transduced to electrical signals, are then inverse-filtered to remove all but the varying amplitude of the frequency at which the system is resonating. This variation contains the signal to be detected. In this case, damping is deliberately kept to a minimum.

In many applications, however, the aim is to introduce damping deliberately to produce a certain response to input excitation. This is done by adding damping to the resonant system with suitable mechanical, or other regime, dissipation elements. Consideration of the means to reduce the effects of mechanical hysteresis suggests that the addition of damping is not always the best approach to use, because vibration to rest can often improve the static precision of such elements as bellows, springs, and diaphragms.

Thorough study of the dynamic performance of an instrument is far more complex and time consuming than is the case for the relatively simple static domain. Certain mechanical parts may not be stiff enough or may possess too much mass when under the rapidly changing movements existing in velocity, acceleration, and jerk conditions. The first step to designing for the dynamic mode in a machine linkage begins with studies that provide the designer with maximum values of velocity, acceleration, and jerk. The kinematics of linkages is a topic well covered in texts on the design of mechanism, see, for instance, Beggs (1955) and Mabie and Ocvirk (1958). This leads to the stress levels that occur, from which the deflections can be calculated (although often with considerable difficulty, for the maximum deflection and dimensional change points are not always in obvious positions). Inertia force considerations calculated from knowledge of the moments of sections of components, coupled with the velocity and acceleration data, lead to other dynamic considerations.

Systems modelling

The above description of the static and dynamic aspects of instruments is not based upon an overall unified systematic understanding of the topics but is typical of the ad hoc approach that many designers use. Each of the above observations can be better explained and calculated using a common fundamental mathematical basis for dealing with the performance of physical systems. The mathematical description of mechanical linear systems is the same as that for electrical, thermal, and fluid regimes. By the use of appropriate coefficients, the performance equations of one regime can be interfaced with those of another. Figure 3.6 is an example showing how the moving coil meter can be expressed in a common electrical regime, which can then be studied as an electric circuit to obtain its dynamic behavior. This means that electromechanical systems can be represented by a single mathematical expression. The actual performance will be closely as predicted from this model (in so far as the real situation matches up with the model used).

Today this understanding is represented in the systems approach to physical systems — see Finkelstein (1977) for its use in measurement. This is not to be confused with the much more widely applied 'general systems theory' that attempts to take in far more experiences, including such topics as biological and sociological phenomena. The latter is similar in concept but has to operate without the same degree of mathematical rigor present in the former.

The systems approach for physical systems has been steadily developing from roots emanating from the early radio era when microphones and loudspeakers were first being developed. Various papers were written on what were later seen to be constituent parts of this integrated topic. Eventually the basics were assembled together as a single methodology in such books as Olson (1943), where the 'analogies' were presented as a systematic approach to the design of linear dynamical systems. Initially the approach was developed for the design of electromechanico-acoustic devices. It gradually found its way into the description of electromagnetical machines in general — see Fitzgerald and Kingsley (1961) and Koenig and Blackwell (1961). The material was later presented as a method to cover the dynamics of all physical system regimes, Shearer, Murphy and Richardson (1971) being a well presented and extensive work; it also lists other books of relevance.

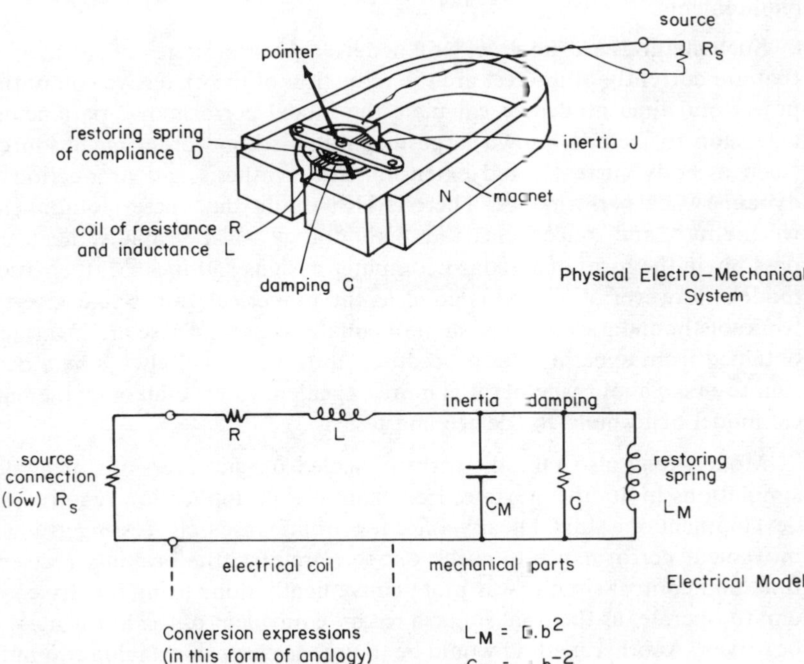

Figure 3.6 Unified representational model, made in the electrical regime, of electro-mechanical moving coil indicator

Each of those above mentioned works was in general written for study of energy conversion devices. Instruments are a class of those specifically concerned with conversion of information into knowledge, that is, using an energy conversion transmission path to convey meaning between points in a system. The foregoing theory of physical systems can be condensed somewhat when instruments are the prime interest. After publishing, with colleagues, several internal reports and conference papers on the topic, Finkelstein and Watts (1978) of the City University, London, presented a systematic view of mathematical modelling of instruments in particular, thus helping the instrument designer to confine his area of study to the matter that is directly of use.

As well as ordering the systems knowledge of instrument transducers, the above mentioned group have used the method to design transducers along rigorous mathematical lines. They have shown here, and later elsewhere (see Abdullah, 1977 and Hugill, 1978a), that it is possible to prepare a paperwork design for a wide range of devices — pressure bellows, differential pressure cell units, inductive gages, capacitive sensors, load cells — that, when manufactured, provides performance that is very close to design requirements.

Such methods are, however, still in developmental stages, problem areas that are currently of interest are the reduction of the extensive computing power and time needed to calculate the model performance parameters, extension to include known large magnitude second-order error sources (such as eddy currents) and extension into synthesis and production of dynamic state performance. There is little doubt that these requirements will be met, and, once done, the methodology will be rapidly taken up, possibly in the form of standard computer designs routines. Mathematical models, however, are only as good as the numerical values that give the workable boundaries to the system of equations used. These are invariably obtained from experimental procedures; thus, there will always be a decision to make about whether it is more expedient to generate a mathematical model or to build the device and test it.

Models may also take the form of scaled physical versions or analog simulations in another regime. For example, an application required the development of a small linear vehicle for robotic research. Testing its linear movement performance to check out the circuitry, the braking, acceleration, and control circuits was more conveniently done using the drive system to operate, as the final stage, a rotary equivalent of the linear mass to be driven. Another method would be to use the electronic analog computer (certainly not made redundant by digital computing), which is often better for studying complex multiparameter control systems (of such things as flight dynamics) than the use of the real object. Digital simulation is also another choice, but not necessarily the best. Each study needs its own consideration to allow for the many factors of that particular study.

A word of caution must be given, however, on the use of models. It is all too easy to become used to the idea that the model that has taken so much hard work to develop is a true and entirely faithful representation of the final instrument. Models are rarely more than just adequate because the effort needed to improve them beyond a certain point is not seen to be further productive. The user decides to compromise in order to make progress.

A model, although it may well simulate the correct input-output relationships, may not do this by the same internal mechanisms as the original system. The field is covered by the discipline of 'systems identification' and 'simulation'. This point is demonstrated by thinking about the fact that there are many technological ways to achieve an end: fuel gages in a car can be electronic or entirely mechanical. Models must be so constructed that any internal data that are desired are available to the user of the model.

The united 'analogies' approach to instrumentation also shows that the performance of an instrument system can be tailored by the use of any appropriate discipline. In setting up an instrument design, the degree of mechanical proportion needs consideration in the light of what can be achieved using alternative electrical, optical, and fluid regimens. For example, several years ago the design of gages for dimensional measurement generally required the designer to incorporate as much mechanical sensitivity as possible in order to obtain an adequate mechanical signal output. Today, the electronic alternative allows virtually unlimited dimensional sensitivity: fractions of nanometers can be routine (Sydenham, 1972). Mechanical stages no longer have that constraint imposed on them; the trading off of sensitivity usually allows higher frequency response.

A second example of the use of alternative regimes arose when a radar tracking turret was assembled and found to have an unexpected resonance in a drive shaft. It was more easily overcome by adding electronic compensation circuitry to the control systems than by replacing the shaft with a stiffer design. Instrument design is no longer the matter for a single approach, nor does electronics always offer the best solution.

This brief introduction to the systems approach adopted within the past two decades is, of course, not restricted to mechanical design aspects alone. Just as real instruments transcend the artificial boundaries by which we have traditionally separated our store of knowledge, so will the necessary theory.

Optimizing the design

Design is the procedure whereby a mass of general facts are gradually sifted to sort out which combination will achieve the specification. There is seldom a unique solution and the designer is constantly required to make choices, knowing that a whole range of other possibilities are then omitted from future considerations. Finklestein (1977) and Bosman (1978) discuss this in relation to instruments.

Periodically within the process there is a need for means by which the choice made can be optimized to give the designer confidence to proceed with the satisfaction that things are going well and that the best use has been made of available factors.

Much has been researched and written about optimization, especially in the case of multiparameter control systems. Optimizing on instrument design can certainly be done, using advanced and extensive procedures to reach better arrangements, but the time and cost available to optimize an instrument design will be very much less than that available for improving, say, the yield of a large throughput chemical plant. In many instances optimization is more likely to be done by manual means, using a small calculator. Families of curves can be plotted to allow visual processing of a number of related parameters in order to choose the optimum arrangement.

In instrument design, as elsewhere, optimization requires the definition of what should be optimized — this invariably leads to a situation where all requirements cannot be optimized at the same time.

There is no totally objective method to decide the acceptable tradeoffs. A method has been developed, however, which enables the designer, through the use of 'utility functions', to access each of the candidate designs selected; see Siddall (1972) and Weiner (1976).

Utility functions, usually of initially graphical form, depict to what extent the designer believes the criteria of interest can be related with parameter changes. Figure 3.7 shows six such functions generated by a designer (Hugill, 1978b) to represent his personal assessment of utility of his choice of six important parameters for an exploration gravimeter. This is the first step in the judgment process. Each utility function represented is then assigned a weighting factor, which ranks the critera in relative importance — again as the designer sees it.

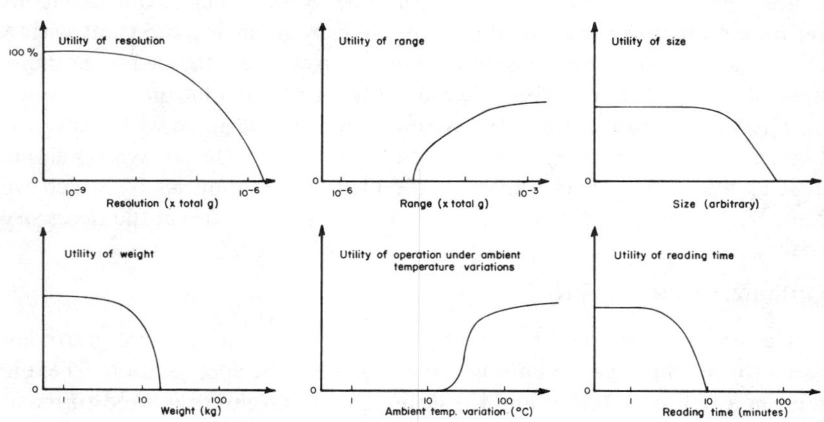

Figure 3.7 Utility of design characteristics for exploration gravimeter — as assessed by Hugill (1978b)

The final stage is to see how each candidate's design rates in terms of the utility functions and their weightings. The final result is a number (the utility value multiplied by the weighting factor for each variable, the whole six results being summed). The highest value represents the best design choice. The other numbers also assist recognition of the degree of superiority of any choice over the others. This is but one variation of the theme expressed by this particular procedure.

It can validly argued that this is hardly an objective method, for each utility function and weighting factor is decided by subjective judgment. Nevertheless, it does provide standards against which subsequent experience can be compared, and it does prevent subjectivity continuing right through the process of choice. One example of the use of subjective criteria is to be found in the proven phase and gain margin values used to condition a linear control loop.

Computer-aided optimization can certainly be used in instrument design but its cost effectiveness is still rather limited because of the lack of developed programs. More work is yet to be done on these, in such places as tertiary educational institutions, before they are in widespread use. There are, however, some useful relevant computer-aided developments already in existance.

The first example is a group of programs developed at the University of Reading, England, to help a designer to condition a closed-loop controller. Basically the computer takes over all systematic rigorous and objective functions of processes like the Nyquist stability assessment procedure, the designer making the subjective decisions such as adequate gain and phase margins, overshoot, rise time, and the like. It is reported in Atkinson (1972). General-purpose widely used needs such as this are worthy of the time spent on them.

Other examples of computer-aided design (CAD for short) that are relevant to instrument systems are those already mentioned of the City University, London, on transducer design and those at Delft Technological University, The Netherlands, which relate to the provision of response of multiregime system networks that the designer specifies and tests with the various forms of excitation. There is little doubt that CAD will gradually penetrate more of instrument design, especially as the general design of instruments gives way to more rigorous procedures.

The need for optimization and efficient control of design does not stop at the stand-alone instrument. It must extend into systems of instruments — the so-called extensive system — and into the role that it plays throughout its working life.

The subject of extensive design is summarized in Bosman (1978). Methodology is presented there that enables the less objective aspects of large systems to be put into an objective framework of design control. In many ways this paper compliments those by Finkelstein previously referred to above, but it goes beyond the stand-alone instrument level. Sydenham

(1979) presents discussion of other aspects such as reasons to improve design, costs of measurement, and how more rigor can be introduced.

All systems are prone to error from the various tolerances of their parts or from external disturbances. Good design aims to ensure that these are adequately controlled so that the worst case is still adequate.

Degradation of performance from that intended can occur because the tolerances of the combined individual components occur out of range (in which case a production batch will have a few defects), because of variation of the tolerances of the many parameters with time (in which case a working instrument will go out of specification after working adequately) or because of the effect of external influencing disturbances (which can cause internal tolerances to be altered while an instrument is operating). Models of the various kinds enable these effects to be studied by the designer.

In sensitivity analysis the basic procedure is to vary each known dominant parameter, one by one, systematically, holding the others constant. This can be done on mathematical models as well as physical ones. Techniques are available to compound the results of these tests. This kind of study can be extremely time consuming because even the small systems, such as are found in small instrument stages, can be influenced by many parameters. Clearly computer aids are useful here.

The 'operating surface' concept developed by Stein and reported in many papers by him (such as Stein, 1971) is a method for graphically presenting the relationships between inputs and the output of an instrument stage. Figure 3.8 shows the operating relationships between inputs and outputs of a system. It can be used to choose the best operating point, to assess linearity, and to visualize how the system responds to overload. By assuming that one or more inputs are noise sources, the same procedure helps the designer to appreciate their effect and where to best operate the system.

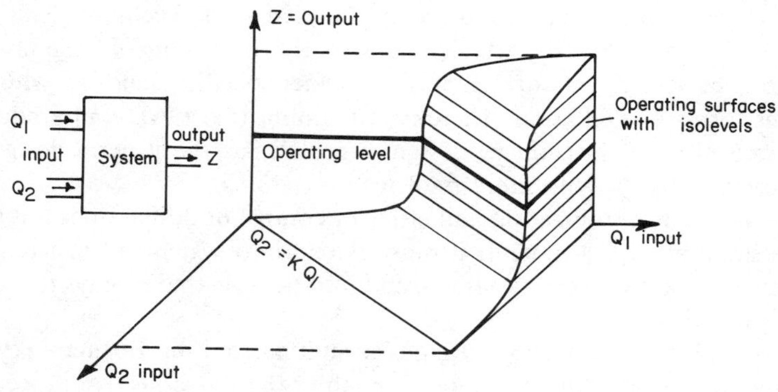

Figure 3.8 Desirable operating surface for a system in which a reference balance is used between inputs Q_1, Q_2 for output Z

Another aspect that might be classified as optimizing a design is the reliability of a system. Well developed methodology now exists by which parameters such a 'mean time between failure' (MTFB) and 'mean time to repair' (MTTR) for individual subsystem components can be compounded to realize the total system reliability. The reader is referred to Gilmore (1969), Simpson (1976) and Breuer and Friedman (1977); although primarily aimed at electronic systems, the concepts apply to mechanical arrangements. Hahn and Shapiro (1967) is also relevant, being on statistical models.

This introduction to the many systems facets of the mechanical design of instruments has shown that mechanical design today can be considerably more rigorous and efficient than in past decades. Most of the concepts described here have been evolved and introduced into general use since the 1940's heyday of the design of fine mechanisms. They have not yet been brought together in any text on the mechanical design of instruments. Work that will assist the reader to gain more references and deeper discussion are the bibliography prepared by IMEKO (1980) and the multicontributor handbook edited by Sydenham (1982, 1983).

Techniques of design that are available today contrast strongly with the methodology described by Whitehead (1934). It must be admitted, however, that many instruments are still being designed by much less systematic and rigorous means than are now possible because of the lack of suitable guidance and infusion of these methods into current training and into the appropriate literature.

References

Abdullah, F. 1977. The application of mathematical models in the evaluation and design of electromechanical instrument transducers. *J Appl Sci Eng*, A, 3—26.

Antill, J. M. 1974. *Civil engineering management*, Angus and Robertson, Sydney. (see Engineering contracts).

Atkinson, P. 1972. 'Computer-aided design of closed-loop control systems', *Computer Aided Design*, April, 120—128.

Beggs, J. S. 1975. *Mechanism*, McGraw-Hill, New York.

Bell and Howell. 1974. *The Bell and Howell pressure transducer handbook*, CEC/Instruments Div, Pasadena.

Bosman, D. 1978. 'Systematic design of instrumentation system', *J Phys E: Sci Instrum*, **11,** 97—105.

Braddick, H. J. J. 1954. *The physics of experimental method*, Chapman and Hall, London.

Breuer, M. A., and Friedman, A. D. 1977. *Diagnosis and reliable design of digital systems*, Pitman, London.

Burstall, A. F. 1970. *A history of mechanical engineering*. Faber and Faber, London.

CAMA. 1974. *A guide to the procurement of complex electronic control and supervisory systems*, The Control and Automation Manufacturers Assoc, London.

Cambridge Instrument Co. 1945. '50 years of scientific instrument manufacture', *Engineering*, **159,** 361–363; 401–403; 461–463; 501–502.

Doebelin, E. O. 1975. *Measurement systems: application and design*, McGraw-Hill Book Co, New York.

Faires, V. M. 1955. *Design of machine elements*. Macmillan, New York.

Finkelstein, L. 1977. 'Instrument science — introductory article', *J Phys E: Sci Instrum*, **10,** 566–572.

Finkelstein, L., and Watts, R. D. 1978. 'Mathematical models of instruments — fundamental principles' *J Phys E: Sci Instrum*, **11,** 841–855.

Fitzgerald, A. E., and Kingsley, G. Jr. 1961. *Electric machinery — the dynamics and statics of electromechanical energy conversion*, McGraw-Hill, New York.

Foxboro. 1972. *Process control instrumentation*. Pub 10SE, Foxboro-Yoxall, Surrey.

Gilmore, H. L. (1969) *Integrated product testing and evaluation; a system approach to improved reliability and quality*, Wiley, New York.

Glazebrook, R. 1922. *A dictionary of applied physics*, 5 Vols, Macmillan, London.

Hahn, G. J., and Shapiro, S. S. 1967. *Statistical models in engineering*, Wiley, New York.

Hugill, A. L. 1978a. 'Synthesis of inductive displacement measuring system using computer-aided design', *Proc IEE*, **125,** (5), 417–421.

Hugill, A. L. 1978b. *Gravimeter design*, Internal Report DSS/AH/171, Systems Science Dept, The City University, London, September.

IMEKO. 1980. *Bibliography of books published on measurement science and technology — with emphasis on the physical sciences*, IMEKO, Higher Education Committee.

Koenig, H. E., and Blackwell, W. A. 1961. *Electromechanical systems theory*, McGraw-Hill, New York.

Lent, D. 1970. *Analysis and design of mechanisms*, Prentice Hall, Englewood Cliffs.

Mabie, H. H., and Ocvirk, F. W. 1958. *Mechanisms and dynamics of machinery*, Wiley, New York.

Moss, C. 1978. 'Quality of precision measurement equipment', *NCSL Newsletter*, **18,** (1), 23–24.

Olson, H. F. 1943. *Dynamical analogies*, Van Nostrand, New York.

Rolt, F. H. 1929. *Gauges and fine measurement*, 2 Vols, Macmillan, London.

Rothwell, E. 1966. 'The characteristics of the printed motor', *Control*, **10,** (93), 136–139.

Shearer, J. L., Murphy, A. T., and Richardson, H. H. 1971. *Introduction to system dynamics*, Addison-Wesley Pub Co, Reading.

Sidall, J. N. 1972. Analytical decision making in engineering design, Prentice Hall, New Jersey.

Simpson, A. 1976. *Testing methods and reliability (electronics)*, Macmillan, London.

Stein, P. K. 1971. 'The concept of a system 'operating surface' as a fundamental system characteristic', *IEEE Trans-IM*, November, 183–191.

Strong, J. 1938. *Modern physical laboratory practice*, Blackie and Son, Glasgow.

Sydenham, P. H. 1969. *Manual and automatic large-scale metrology*, PhD Thesis, University of Warwick, England.

Sydenham, P. H. 1972. 'Microdisplacement transducers', *J Phys E: Sci Instrum*, **5**, 721–733.

Sydenham, P. H. 1979. *The systematic design of measuring systems*, Proc Engineering Measurement Systems Conf, ACIS, Wellington, New Zealand, May.

Sydenham, P. H. 1982, 1983. (Editor) *Handbook of measurement-science*, John Wiley, Chichester.

West, J. C., and Leonard, J. L. 1955. 'The necessary torque requirements for a servo-motor', *Jnl Phys E: Sci Instrum*, **32**, 30–32.

Wheeldon, R. 1974. *Specifications — the identifying facts and Generalized approach to selection and design (of instruments), in Stathmology — the integrated approach to measurement, instrumentation and control*, Department of Continuing Education, University of New England, Australia.

Whitehead, T. N. 1934. *Instruments and accurate mechanisms — underlying principles*, Macmillan, London (Also Dover Pubs, New York, 1954).

Weiner, Ul. E. 1976. *The value of measuring information, determination possibilities, application of management analysis of measuring systems*, IMEKO VII Congress, paper CTH / 303, May London, IMEKO Secretariat, Budapest.

Chapter 4
Kinematic components

Introduction

Having dealt with the general aspects of instruments in the previous chapter, attention can now be given to the many individual mechanisms that are used together to create a required mechanical system.

Mechanisms act as links, called pairs, that couple parts of a system together. Although not a classification generally used, it is convenient (and logical) to divide mechanisms into two groups those based mainly on kinematic concepts and those that operate via elasticity. For our purposes, links that are strictly semikinematic because of their bearing area being greater than a point contact, including those which are manufactured to perfect enough tolerances, will be placed in the kinematic class forming the first group. Both types do use elasticity to spread the initial load but do not rely on compliance to allow subsequent deformations to take place. For example, a bolted joint deforms as it is tightened but does not flex as would an elastic cantilever beam that it could be holding. Ball bearings also deform at the points of contact, but the degree of yield is generally smaller than the component they might be supporting.

As a general rule, this and the following chapter on elastic links will give greater emphasis to the instrument aspects not commonly discussed in machine element design texts.

The various kinds of mechanism that might be employed include such devices as members, columns, bearings, linkage arrangements, wobble pins, gears, belts, cams, fasteners, flexures, shaft couplings, and clutches. There are very many of these, and in the past people have always attempted to classify these into logical groups and form representative models of them. Today, it seems from the published material that no satisfactory classification has been achieved. Authors still differ widely in the way that their works are organized.

In this account they will be ordered to reflect the design process: how the designer moves from one stage to the next, gradually completing the detail of the system being created. (In practice, however, the design process involves many stages of iteration between the steps of choice of mechanism to actual realization).

Here kinematic and semikinematic mechanisms are divided into members, linkage arrangements of members, fastenings that join members rigidly, bearings that are used at the joining points of linkages to allow motion, and the transference of motion across a linking point where rotation is required at a given gearing ratio.

Three relevant tables are given. These are: collapsing limits of columns and shafts; moments of section; stress and deflection of beams and similar members.

Texts that contain information on the various types of mechanisms include Beggs (1955), Braddick (1954) for which a 1966 edition exists, Faires (1955), Mabie and Ocvirk (1958), which was reprinted in 1975, Peatfield (1951), Doughtie (1966), Reliance Gear (1972 and other years) and Trylinski (1971). The many publications of Geary (see references of Chapter 1) may also be of use in the general selection of mechanism. Begg's work contains a 'repertory of mechanism' that is provided as a source of ideas. Of these books, Trylinski's work is the most theoretical general work that is specifically on instrument mechanism. The date of publication of works on mechanism is not very significant for mechanisms have been known for centuries. Very few new mechanisms, in fact, have been invented in these past few decades. A paper by Furse (1981) is relevant reading.

Members

In this subgroup the discussion centers on three-dimensional materials that are fashioned to have significant length over radius of gyration ratios. Included are columns, beams, plates, and shafts.

A slender member can be used in tension with the only failure mechanism being when it yields beyond its elastic limit. The elasticity and the strength are calculated from the usual strength expressions relating Young's modulus of elasticity to compliance and the material parameters that define the limits of tensile load. Compliance remains the same for all conditions of a material; tempering and other treatment does not alter the elasticity, only the ultimate point at which yielding begins and Hookean elasticity ceases. The exception to this rule is when the material is formed as an extremely slender fiber; in that case the properties change significantly. It is well to remember that the end connections may need to be of much larger section than the member to keep bending stresses low. In instrument design, where dimensional change is important, the compliance of apparently adequately strong members may be too high. Calculations should be made to verify that it is adequate.

In the tensile member the forces act to stretch the member; in this case the force system is stable. If, however, the force is reversed so that it compresses the slender member, the failure that may first appear might not be compressive yield but may be due to the column buckling as it bends to form one of several possible modal shapes.

This form of collapse is quantified by use of the Euler's column formula (Timoshenko and MacCullough, 1959). The formula gives the value of load at which the fundamental mode, for the given end connections, begins to appear. Buckling has been used as a form of high-rate spring, see Chapter 5.

Euler's point of buckling depends upon the end conditions of constraint that are applied to the two ends of the beam. These may be so-called 'pin-connected' (that allows rotation at the end) or be firmly clamped so that rotation is not possible at the point of connection. The first case is also referred to as 'simply-supported', the latter as 'built-in' or 'fixed-ends'. Another condition of Euler buckling can also occur when the end is not constrained in translation, being free to move sideways under the load.

An interesting feature about the two sets of beam formulae (Euler and deflection) is that although each has different expressions, within each group the expressions are identical except for their numerical coefficient. The stiffness of a column or beam increases as the ends are made more constrained. Table 4.1 lists the various cases of collapse. As a rough guide, Euler collapse begins to occur when the slenderness ratio rises to be around 100. The value must be calculated for the smallest radius of gyration when the member has an asymmetric cross section. Table 4.2 provides information about the moments of typical sections of structural member.

Table 4.1
Columns and shafts — collapse point

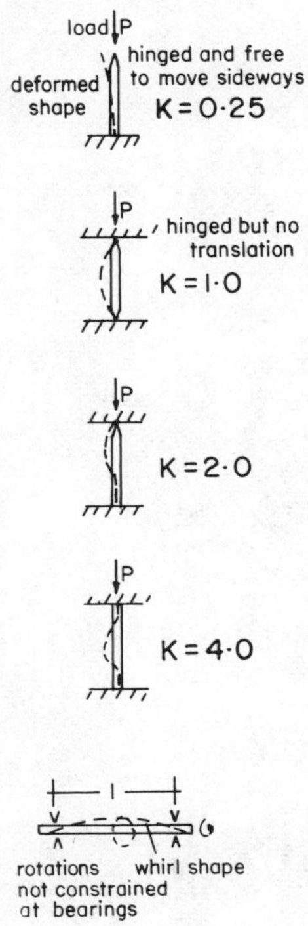

columns

general expression for least strength mode of supportable maximum load

$$P\ critical = K.\frac{\pi^2 E I}{l^2}$$

where P cr — initial load at point of collapse

E — elasticity modulus

I — least second moment of cross-section

l — length between ends

shafts

lowest speed at which whirl begins is

$$N = \frac{\pi}{2l^2}\sqrt{\frac{EI}{w}}$$

where E – elasticity modulus I – second moment of cross-section

w – mass per unit length of shaft

l – distance between bearings

Rotating shafts also have a performance failure that can be likened to the column collapse characteristic. A shaft that is free to turn in bearings that do not constrain its rotation at the bearings will reach a critical speed at which it begins to whirl — the central region attempts to move outwards due to centrifugal forces. It takes up a bent profile while it rotates. This, too, depends upon the elasticity of the material and the slenderness ratio of the shaft. The expression giving the lowest whirling speed is given in Table 4.1. Attachment of loads along the shaft alter this value (Faires, 1955, and Mabie and Ocvirk, 1958).

Table 4.2
Second moments of commonly met sections (Composites can be built up by addition of these)

	Section	Moment of Inertia	Radius of Gyration
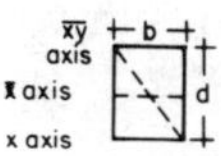	Rectangle	$I_{\bar{x}} = \frac{bd^3}{12}$ $I_x = \frac{bd^3}{3}$ $I_{\bar{x}\bar{y}} = \frac{b^2d^2}{4}$	$k_{\bar{x}} = \frac{d}{\sqrt{12}}$ $k_x = \frac{h}{\sqrt{3}}$
	Circle	$I_{\bar{x}} = \frac{\pi r^4}{4}$	$k_{\bar{x}} = \frac{r}{2}$
	Tube	$I = \frac{\pi(r_1^4 - r_2^4)}{4}$	
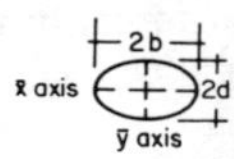	Ellipse	$I_{\bar{x}} = \frac{\pi bd^3}{4}$ $I_{\bar{y}} = \frac{\pi b^3 d}{4}$	$k_{\bar{x}} = \frac{d}{2}$ $k_{\bar{y}} = \frac{b}{2}$
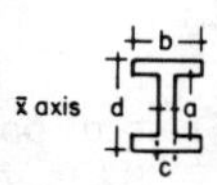	I-section	$I_{\bar{x}} = \frac{[bd^3 - (b-c)a^3]}{12}$	
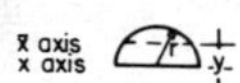	Semicircle	$I_x = \frac{\pi r^4}{8}$ $I_{\bar{x}} = 0{\cdot}11 r^4$ (where $\bar{y} = \frac{4r}{3\pi}$)	$k_x = \frac{r}{2}$ $k_{\bar{x}} = 0{\cdot}264r$
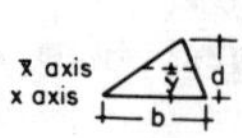	any Triangle	$I_x = \frac{bd^3}{12}$ $I_{\bar{x}} = \frac{bd^3}{36}$ (where $\bar{y} = \frac{d}{3}$)	$k_x = \frac{h}{\sqrt{6}}$ $k_{\bar{x}} = \frac{h}{\sqrt{18}}$
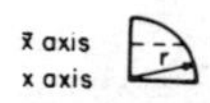	Quarter circle	$I_x = \frac{\pi r^4}{16}$ $I_{\bar{x}} = 0{\cdot}055 r^4$	$k_x = \frac{r}{2}$ $k_{\bar{x}} = 0{\cdot}264r$
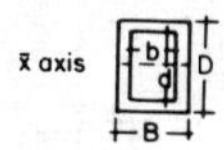	Box section	$I_{\bar{x}} = \frac{BD^3 - bd^3}{12}$	$k_{\bar{x}} = \sqrt{\frac{BD^3 - bd^3}{12(BD - bd)}}$
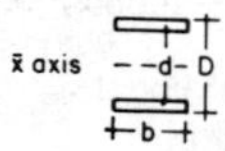	Two beams	$I_{\bar{x}} = \frac{b(D^3 - d^3)}{12}$	$k_{\bar{x}} = \sqrt{\frac{D^3 - d^3}{12(D-d)}}$

In general $I = Ak^2$ where A is area of cross section and k is radius of gyration

Slender members might be used as shafts coupling motion, or as parts of a linkage to produce a given motion. They can also form wobblepins that transfer linear motion without forcing any constraint on the freedom of translation that occurs at the end of the members forming either side of the drive. Wobble pins, Figure 4.1, can clearly only be used to transmit compressive force, since the force keeps the point contacts in connection. Ends are usually formed to be small radii hemispheres. Another use for slender members was given in Chapter 3 where thin legs were used to support a tiltmeter while allowing freedom of expansion of its body. If in tension, drive can be transmitted by relatively thin wires or rods, for there the Euler collapsing mode cannot occur.

Slender members (often called beams) used in the horizontal direction to support loads are the usual subject of texts on the strength of materials; origins are largely for civil engineering structures analysis. The various cases of point load at a given position along the beam, a uniformly distributed load or of a mixture of loads are each well developed for the various free and clamped-end connections.

The formulation extends to cover the two-dimensional plate having freely supported and built-in edges to which is applied a central point load or a total area distributed load. These latter cases cover the needs of most diaphragms in pressure gages. Deflection of shaped diaphragms is discussed in Trylinski (1971).

In instrument design it is usually the transverse deflection of beams rather than the stresses that will be of the most importance. A table listing the most commonly needed cases is provided in Chapter 5. Many combinations of instrument designs tend to use the simpler cases. It is of some value to see that for the beam supporting a point load the formulation is the same for the various end connections, except for the constant, which varies for each case. A similiar situation also applies to the distributedly loaded beam. Distributed loads result in less transverse deflection than for the same gross point load case. The more rigid the end points, the smaller the deflection. It

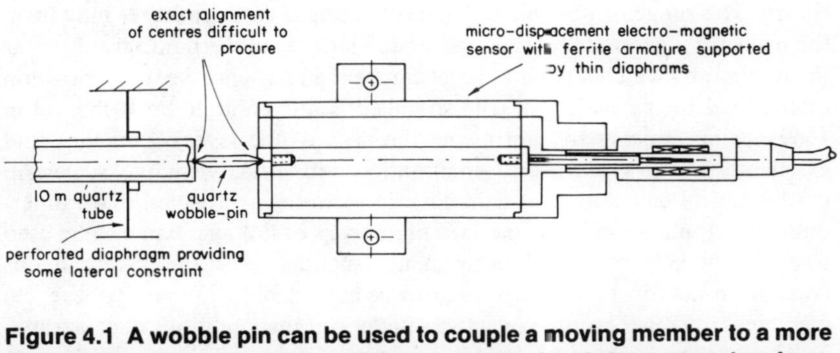

Figure 4.1 A wobble pin can be used to couple a moving member to a more constrained point. It will buckle if too slender for the compression force (Sydenham, 1974).

will generally be found that if the deflection is realistic, the stress level will be well below the yield point. Some spring designs, of course, will break this general rule. When using beams as rigid members, it is wise to calculate the likely deflection where dimensional change is important.

No difficulty should be experienced in obtaining the various formulations. In the case of very complex loading it may be necessary to develop the expression from the basic approaches that are used to obtain the quoted equations; see Timoshenko and MacCullough (1959) for a full treatment of the equations of beams and other components. Texts containing tables giving the common expressions are Braddick (1954), Faires (1955), Neubert (1975), and Timoshenko and MacCullough (1959).

Beams that are curved and have to carry linear tensile or shear loads, such as a hook, will experience high bending stresses in the region of the bend. Areas around rivet and pin connections can be unexpectedly weak near the joint if care is not taken to increase the section size.

Beams find use in instruments as structural members and as sensing members; see the next Chapter on elastic mechanism for the latter. They may, therefore, need to be inelastic or elastic. Successful design rests on choice of material, design of the member shape and size, and the connections used.

Shafts also can be made elastic or be rigid depending upon the need. They should be large in section and short in length to increase the frequency of resonance when they transfer rotary motion in closed-loop systems. The torsional elasticity of a shaft is calculated from the torsional or shear modulus for the material, the process being analogous to linear elasticity of a member. Dynamic balancing of rotating parts is needed where vibration levels must be kept low.

Linkages

Members connected together can be used to provide conversion of motion from linear to rotary, from linear to linear, and from rotary to rotary. The range of possible linkage functions is vast. Linkages may form the basis for a geometrical size conversion such as in the pantograph, or as the method by which a balance scale can be made insensitive to the position of the load in the scale pan (the so-called 'static enigma' by Roberval in 1669, which appears to contravene the laws of the lever). Linkages find extensive use in typewriters, pneumatic controllers, weighing scales, in mathematical calculating instruments, in recorder pen drives, and numerous other applications. The method of analysis of linkages can also be used to study the action of solid components, such as cams; the instantaneous contact points form geometry that can be considered to form hypothetical members. Linkages become structures under certain conditions of constraint.

The mathematical analysis of linkages has fascinated academics for centuries. Many of the defined fundamental problems are still to be solved in theory. The rise and development of the kinematics of linkage is described in Ferguson (1962). To a large extent, analysis, that method of breaking down the action of a given linkage, has succumbed to methodology that can be used to ascertain the positional motion of linkages of reasonably simple nature. As the number of members in the linkage rises, the problem becomes more difficult. The four-bar linkage begins to present a significant theoretical challenge.

As there does not seem to be a simple general theory that can be applied to decide the motion at the various parts of a linkage system, the problem is often tackled by the use of graphical procedures. To assist the designer, several authors over the centuries have published catalogs that show the motion of variously constrained linkages.

What designers would rather have is the means systematically to synthesize a linkage system in order to obtain a required motional relationship. This is, however, still in development; only relatively simple cases can be handled. Computer solution of the equations involved has extended the use of synthesis methods. Usually the designer resorts to the use of models (in the computer or real), makes use of the above mentioned catalogs, or iterates toward a suitable linkage arrangement using analysis of each choice.

It is sobering to realize that the theoreticians sought for many years the theory by which a perfect linear action can be produced by some point in a system of links. Out of this and similiar interests has arisen much useful mathematics for other disciplines (such as the Chebyshev functions now used in electronic filter design). This linkage was finally produced by Peaucellier (Figure 4.2a) in the 1870's. Many useful systems for instrument work have been invented. The 'static-enigma' balance mechanism made simple-to-use, accurate, top access scales feasible (Figure 4.2b, c). The 'Sylvester—Kempe translating linkage' (Figure 4.2d) allows one platform to move parallel and equidistant from a second surface. Johnson's 'converting motion' provides two revolutions from a single reciprocating translation. Other systems related directly to linkages are the Geneva stepping mechanism, the Scotch yoke, the Oldham coupling, and many more.

Ability to analyze a linkage might be necessary to establish the clearances needed as the members move, to assess the velocities and accelerations of the various parts and points, to predict the input—output functions, and to study the action of systems like cams and gears. The methods of analysis of linkages are discussed in Beggs (1955), Faires (1955), Mabie and Ocvirk (1958), Zimmerman (1962), and Lent (1970). Trylinski (1971) uses the procedures in his discussions of the kinematics of various mechanisms used in fine mechanics. Ferguson (1962) includes several references to linkage theory. The journal *Precision Engineering* (first issue, January 1979) is a source of papers. Burstall (1970) also discusses the history of linkage development giving several references to early works. This book contains illustrations of linkages that were invented during the history of man's involvement with technology. Pollard (1929), on the design of couplings, also might help understanding of the procedures.

Fastenings

The purpose of fastenings is to join components together in a manner that will provide firm constraint of all degrees of freedom at the joint. Existing designs vary greatly, the choice largely depending upon the magnitude of the forces and torques involved, whether the joint is permanent or

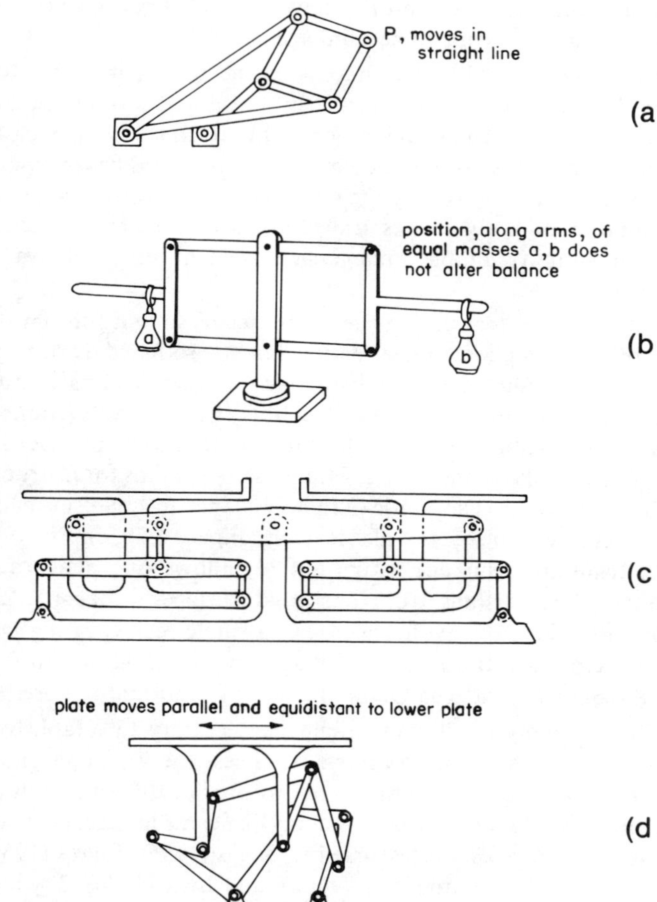

Figure 4.2 Some linkages used in instruments (a) Peaucellier's straight-line generator — one form; (b) Roberval's static enigma; (c) Beranger's use of (b) in a scale; (d) Sylvester-Kempe parallel translating linkage.

needs to be detachable, the degree of vibration experienced, and the level of dimensional accuracy needed. The design of the joints is equally as important as the other factors of a mechanical system.

An introductory, nonmathematical catalog of fastenings can be found in Peatfield (1951). A thorough treatment of the many types possible is given in Trylinski (1971), this being a key work of reference. Machine design texts such as Faires (1955) and Lent (1970) generally treat only the joints used in power machines, that is, bolted, riveted, and welded arrangements.

In some applications the dimensional stability of the joint can be very important. One view suggests that in such cases it is better to avoid using joints, making the whole unit from a single piece of material. Examples are the manufacture of flexure pivots in which the flexure part is formed by thinning down the piece of material. There is, however, very little substantive evidence to prove that this is necessary. To the contrary, it has been found that remarkably crude joints such as knotted metal wires, rings placed over a shaft and in tension, small magnets holding a part to a surface and bolted joints, (Jeffery and Sydenham, 1973) will each hold dimensional relationships across the joint to less than micrometer level. Couplings working at the very small tolerances of nanometers or less may well exhibit instability, but it is far from easy to verify where such drift actually occurs. Errors of this magnitude can come from so many sources; see the discussion in later chapters. As an example of the many forms of spring-gravimeter that have been made, units using bolted metal arrangements have shown as small a drift rate as others with mostly continuous parts.

Joints forming part of a mechanical resonance loop can, however, seriously alter the Q-factor because of damping inherent in the jointed faces. Joints should be avoided where high quality factors are needed.

In instruments, the joints must provide adequate strength but also they must usually be stiff in compliance. Due to the small sizes of the metal work in instruments, many ingenius joints can be used that would not be practical for large-size machine construction. Joints divide into permanent and detachable methods.

Permanent Joints

Figure 4.3 illustrates examples of some methods of making a permanent joint. A commonly used permanent joint is formed by passing the metal of one part through a hole in the other, swaging or twisting it to lock the two parts together. The tang can be twisted, bent over, swaged flat, center punched, or crimped. Trylinski (1971) gives illustrations of many types.

Force-fit joints that rely on friction and a certain degree of cold welding as the parts are forced into unison can also be used. For these the tolerances of the mating parts must be carefully controlled. These give good dimensional control.

Welded joints are often used, the smallest possible area of welding being at micrometer size. Fine pressure capsules are often welded using focused laser radiation. Gas or electric welding is used. Distortion due to the heating must be allowed for. Not all materials can be welded.

Adhesives are used to form permanent joints, the modern epoxy formulations having considerably changed the role of adhesives in instruments. Work at the Sira Institute in England has resulted in several publications on the use of adhesives, Shields (1976) being a substantial account. The slow setting rate of cements and adhesives can make them unsuitable for

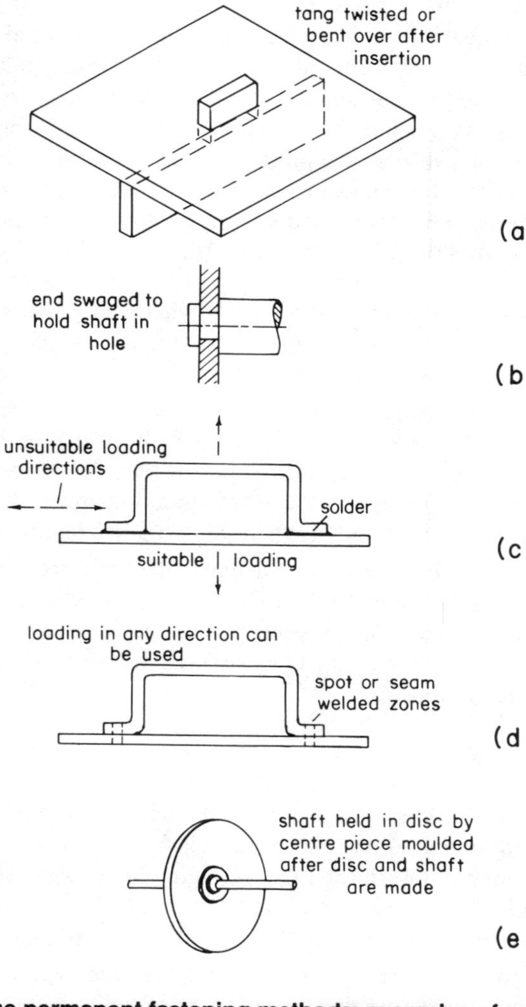

Figure 4.3 Some permanent fastening methods; examples of each type: (a) Bent tangs; (b) Swaged after insertion; (c) Soldered or glued; (d) Resistance welding; (e) Molded in place.

mass production methods. In some cases, cements, those substances that fill a gap without stressing the parts that are held together, may be the most suitable method of securing components, for example, holding a lens in a holder.

When using glued or cemented joints, it is necessary to ascertain if the joint will have adequate heat transfer or electrical conduction across the joining layer that is interposed between the parts. Cements and adhesives are often hygroscopic; they may also creep considerably. In some instances, however, glueing may be the only satisfactory method of making a joint. Good design attempts to form a fillet of glue or a large contact area. The glue should not be used to form a dimensional link but be used to adhere two creep-free materials in close contact. Trylinski (1971) contains a large bibliography on the subject.

Detachable Joints

Figure 4.4 shows some examples of detachable methods of jointing. These can be made using screw threads holding flanges together or which are formed on the end of rods to be jointed; using pins holding a shaft in a hold (tapered, parallel, roll-pin, shear-pin); using keys that hold a shaft in a hole; and as press fit combinations. Where the jointing forces are low, such as the holding of a dust cap over a socket, the use of the lids and caps held by simple turn-lock devices can be used. Trylinski also covers these in detail.

When bolts are used to hold a flange, the designer has the choice of making the bolt hole with close fit tolerance to ensure the correct dimensional relationship or to use some other method of location using the bolt only as a clamping device. Locating methods in use are spigots or taper and parallel cylindrical pins. It is not good practice to use a large clearance hole in which the actual position is set up on assembly. This may suffice for a one-off prototype, but in conditions of vibration or where service personnel may not understand what is needed the position can be lost. Where positional accuracy is vital, the joint can be drilled and pinned after assembly. A spigot is used where the location is predetermined. Where the bolt is used to retain position, its shaft should define the location, not the thread diameter. To ensure that this is done, the thread is made at a smaller outer diameter than the shaft next to the head of the bolt.

Screw threads are used extensively and numerous forms of thread profile and diameter are in use. The designer of a production instrument should strive to use a standard thread where possible. Different disciplines have their own threads — cameras and microscopes being examples. Multiple start, special thread profile, taper threads, and others are sometimes used in instruments. Threads can be locked with some form of lock or tab washer or be pinned. Where the joint must be secured against unauthorized opening, the shaft outside of the nut is drilled to take a small wire that can be sealed. Special proprietory fluids are available to lock threads; paint is sometimes used. Here, again, Trylinski (1971) is a definitive reference work.

Bearings

Joints between two parts that allow motion, such as rotation or sliding in translation, are generally called bearings. Trylinski calls them supports. The relative motion may combine both rotation and translation through the bearing axis. They may be included to limit end-float in a thrust bearing or to allow an arm to turn about a fixed shaft pin. Figure 4.5 shows a representation of bearings that can be used in instruments.

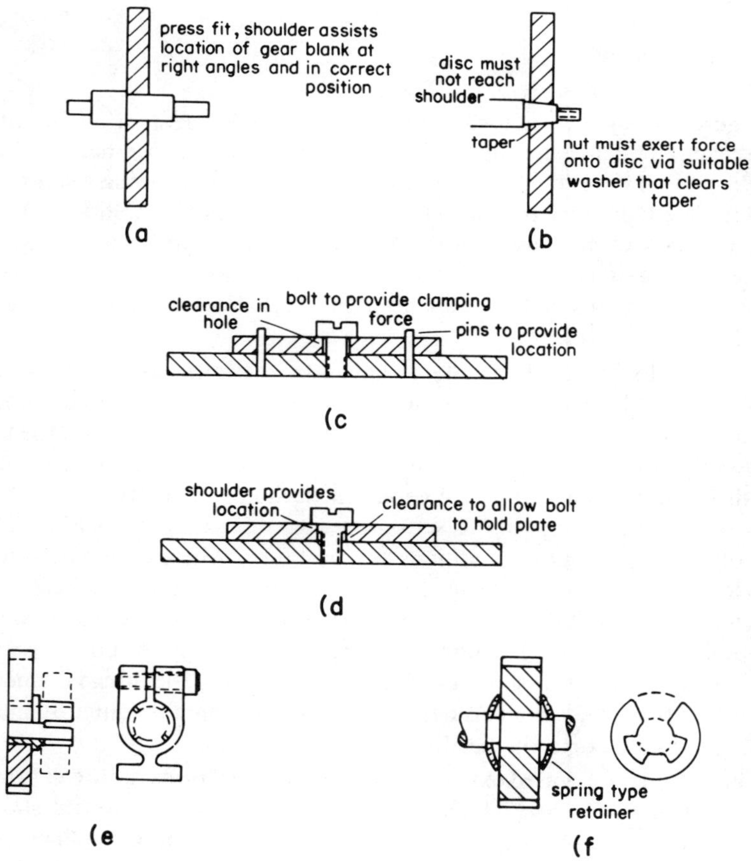

Figure 4.4 Some detachable fastening methods: examples of each type: (a) Press fit; (b) Taper shaft; (c) Pinned; (d) Screw thread; (e) Clamp-hub; (f) Retainer ring

Selection of bearings begins by a definition of the role that the material support has to play. Systems sketches realize the degrees of freedom that are needed; study of the systems operation decides the tolerances that are permissible, for all bearings are imperfect and will add some measure of freedom at the pivot bearing point. Other factors to be considered will be the relative speed of the two surfaces involved, the kind of lubrication that can be used, the coefficient of friction, the life needed, assembly requirements, cost and market availability, effect of temperature changes, and the degree of stiffness needed from the joint formed in the bearing. It is quite possible to produce a usable bearing (using rolling balls) that is as stiff and backlash-free across the joint as the rest of the structure.

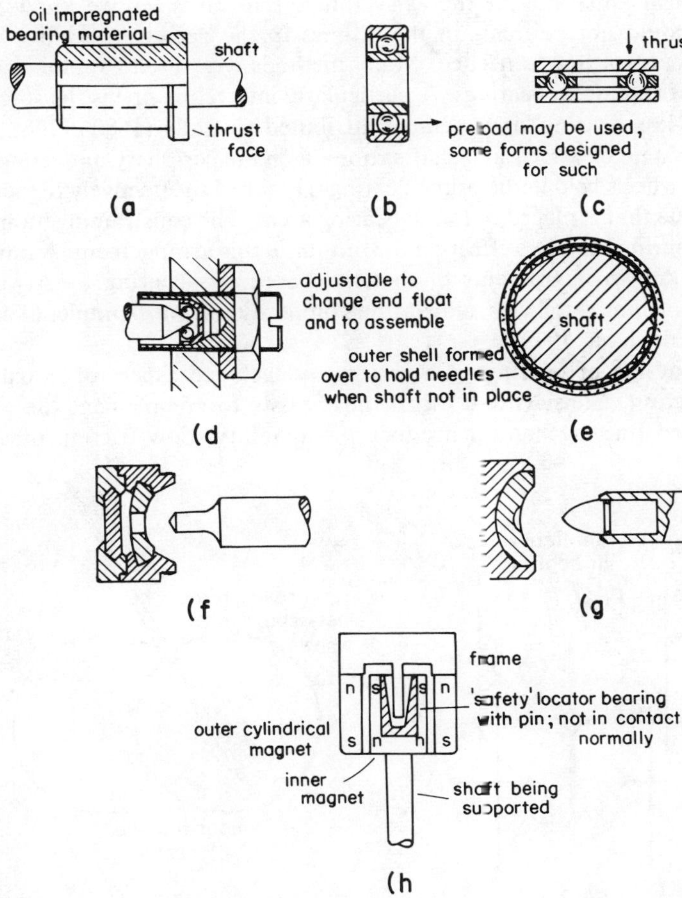

Figure 4.5 Selection of passive bearings used in instruments: (a) Plain bush; (b) Radial ball; (c) Ball thrust race; (d) Pivot type ball; (e) Needle roller; (f) Jewel with thrust; (g) Center support; (h) Magnetic

Supports that allow movement by virtue of elasticity of a flexure are discussed in the following chapter.

Bearings for rotating shafts have been in use since the earliest times. Rawhide was used by the Egyptians around 1500 BC. Bronze bushes were in wagons of Roman times. Crude ball bearings date to the same period. Caligula had a ship with turntables running on large bronze balls. These early ball systems were not strictly freerunning, for they were constrained so that the balls also slid while turning; Burstall (1970) gives more detail.

Around 1770, Count Carburi used a system of metal tracks and balls to move a 1.5 10^6 kg granite mass across the frozen Russian land (Turner, 1973). The tracks were laid end to end as the statue was moved.

Friction was recognized as the limiting phenomenon in many an elegant mechanical apparatus of the early times. Ball races, as we know them today, could not be made in these times to the degree of accuracy and quality control that is needed. Many methods were used to minimize the effects of friction in bearings. A particularly interesting one is the so-called 'friction-less bearing' arrangement attributed to Sully (1680–1728). This used a system of wheels and shafts running on the periphery of overlapping support wheels held in the prime bearings (Figure 4.6) effectively to increase the torque that is placed onto the bearing areas. The constraining moments of the bearing have less effect on the output in this arrangement. A number of well known instruments of that time used this bearing — Atwood's falling machine, Michell's and Nairne's dipneedles being examples (Turner, 1973, Sydenham, 1979).

The design of bearings is a very well developed aspect of instrument construction. Considerable theory now exists to complement the highly developed fine mechanical production capability. Low friction precision

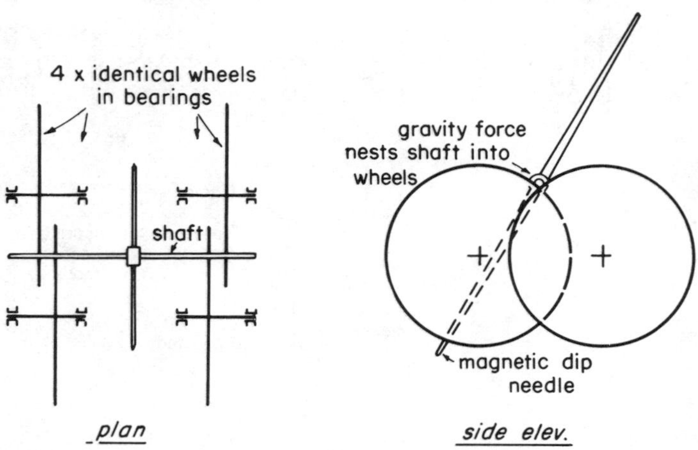

Figure 4.6 The Sully (*c* 1700) frictionless bearing arrangement as used by Michell in a late 18th century dip circle.

bearings can be as small as a millimeter. The designer who wishes to be sure that his choice is justified and within the requirements of the task has a wealth of theory and experimental data to fall back on. Manufacturers provide considerable design material to assist application. Permissable load, run-out, clearances, speed, friction force, life, and assembly arrangements are all well documented.

Since the various forms of bearings are generally familiar, this account will only introduce each kind, giving references to where in-depth information can be found. An excellent work on each type is Trylinski (1971). It provides information about the small and miniature-size bearings that is not generally given in machine-design texts. It covers jewel bearings; direct in the base bearings used in clock mechanisms; conical and tapered supports, which use a tapered shaft end running in a hole in a suitable bush; spherical ends; hydrostatic designs; center supports in which a spherical shaft end runs against a concave plate; cover-plate type thrust bearings, including the shock-resistant Incabloc arrangement; mercury supports; knife edges; rolling-contact bearings, such as the very easy to use cup-and-cone miniature and subminiature ball bearings; and wire track bearings in which the rolling contact ball is used to form its mating track by being pressed into soft metal wires. The chapter on supports covers friction and bearing area stress calculation, errors in location and many other theoretically based explanations of behavior. Eighty-two references are included. A further chapter on guides covers the design of sliding arrangements, which are usually of rectilinear form.

Further works on instrument bearings include the series compiled by Geary of the Sira Institute, (then called the British Scientific Instrument Research Association, BSIRA). His first booklet was on flexure devices and is covered in the following chapter. The second booklet, Geary (1955), was a bibliographic survey of knife-edge bearings. It gives an illustrated technical introduction followed by 166 references. Geary concluded then that knife edges should be considered as blunt or rounded edges, for in practice deformation occurs resulting in a less than ideal knife edge. He also wrote that there was a general dearth of research and objective data available, especially in the English language, and that the best written works were of German origin. He suggested that some should be translated. (Trylinski's work was later translated, helping to fill this gap in available knowledge.)

Geary's fourth report of the series on instrument parts (Geary, 1961), covered instrument ball bearings in a similiar way, providing a specific account on the small series of precision-quality ball bearings. It is a key document where information on the use and choice of small ball bearings is required. Harris (1966) is also relevant.

Applications requiring zero clearance and a large measure of stiffness across the ball-bearing joint can use specially designed ball races — the angular-contact series. In use, the shaft and housing dimensions are chosen to force the balls to ride against one side of the race. This is known as 'preload'. Ball nuts, those drive nuts in which balls run in contact with the lead screw thread, are often used together in pairs, one being rotated against the other to preload the nut into the shaft. Ground leadscrew shafts can provide remarkably stiff, yet free, joints.

For nonmagnetic applications, beryllium-copper ball races can be used. Stainless steel bearings are also sold. Bearings capable of 300 000 rev/min are used in dentist's drills. Each application dictates careful choice from within the various classes of bearings if best operation is to be achieved. The fifth report (Geary, 1962), covered fluid film bearings. This provides a general introduction to this bearing type, gives details of liquid lubricated bearings using external pressure generation, of gas-lubricated bearings, and of less well known bearings such as fluid-squeeze films, mercury globule, and acoustic radiation pressure supports.

The frictional force of the different kinds of bearings varies considerably, ball-bearing systems exhibiting considerably less frictional torque levels than simple liquid-fluid bushes. As an indication of the magnitudes involved, oil-filled gyroscope ball bearings having five balls running at 4 mm diameter might exhibit a breakaway torque of 8×10^{-6} Nm. Running torque is generally about 30 to 50% less. Very small, three-ball pivot bearings breakaway at 1.0×10^{-7} Nm torque. Torque figures increase with load on the pivot. In contrast, a small plain bronze bush needs about 6 to 8×10^{-4} Nm to move from rest. A small radial ball race will need around 3×10^{-5} Nm torque before it begins to rotate. These figures can only be used as a guide — the actual value varies greatly with bearing form, preload, shearload, and shaft size. They must be calculated or measured for the support actually to be used. Misalignment can introduce considerable frictional loss, unless designed as a self-aligning arrangement. Suspensions for electrical indicating instruments use knife edges, jewels, bushes, and elastic bands. Drysdale and Jolley (1924), Part 1, include an extensive discussion of these, most of the material given being about jewel bearings. Geary (1964) covers magnetic and electric suspensions. Permanent magnetic suspensions for instrument use were recently surveyed by Fremery and Doden (1978).

Transference of rotary motion

As in electrical systems, mechanical counterparts require means to transfer signals between subunits of the system. These signals may have to be transmitted at the direct unity ratio or be operated upon to alter this ratio up or down. The motion may need to be converted into such motions as intermittent, quick return, and others.

Transferring signals and forces over large distances in mechanical equipment is more difficult than with electrical signals. Generally, today's practice would now convert mechanical information into electronic for long distance transmission. This was not always so, and there are still applications where direct mechanical transfer is more secure, more reliable, and more easily serviced.

Mathematical uses of mechanisms are covered in more depth in a later chapter on computing mechanisms. Here will be discussed those components that, while not generally being thought of as computing elements, nevertheless might be used in that manner and should be regarded in that light. It is important to see coupling systems in this way for there are many traps to fall into because of imperfections of components and mechanism concepts. Covered here are the shaft coupling devices, gears of the various kinds, cams, flexible belt-type drives, and electromechanical gearboxes.

The form and design of these components are well covered in the general machine design texts such as those already mentioned; Peatfield (1951), Faires (1955), Beggs (1955), Mabie and Ocvirk (1958) and Lent (1970). Information specifically relating to instruments is given in Trylinski (1971) and Reliance Gear (1972). Because of the extensive use of gears and other couplings in industry, there are many national standard specifications.

Couplings

When coupling two shafts together as a direct drive, the factors to be considered are the amount of misalignment that must be tolerated, the constancy of the velocity and positional form transmitted within a revolution, the torque level, the distance between supports to reduce shaft whirl, damping of vibration, stiffness needed in closed-loop control, allowance for end-float and more severe longitudinal movement, ease of assembly, and such factors as availability and cost. A sample of the range is shown in Figure 4.7.

Constant velocity couplings began with the design of Hooke reported in 1674. The single Hooke joint, however, does not give a constant velocity (or position) ratio between the input and the output. Two can be used together in series so that the modulation imposed by one is cancelled by the other. Other more compact forms of constant velocity joints were later developed for the automotive industry, examples being the Bendix-Weiss, Rzeppa, and Tracta joints. These, in general, have not been produced at instrument size. They have the advantage of providing constant velocity from a single coupling joint. The above coupling stages require the two coupled shafts to be positioned such that their extended center lines intersect in the coupling center. A double Hooke joint can, however, also take up translational as well as angular misalignments.

Where two shafts are parallel but not in line, the Oldham coupling can be used. This allows for misalignment by a sliding, cross-shaped connector.

A solid or rigid coupling can only be used where the two shafts are aligned to exacting tolerances. If they are not, the coupling will 'work' on its screws or bend the shafts, causing looseness or failure to occur. Coupling drives come in many forms; see Reliance Gear (1972) and Trylinski (1971) for illustrations of the various instrument kinds. The latter work presents the geometrical theory for many designs. The following names are used in the literature for the various kinds of couplings; slip joint and slip couplings, membrane, spring and self-locking spring, antibacklash, bellows, molded plastic flexible, universal lateral, Kudriavetz, Flex-thane multiple polyurethane pins, multijaw, sleeve, slip clutch, overrunning clutch, and phase adjust (a type in which the relative shaft phase can be slightly adjusted by a small screw rotation). There is no shortage of design possibilities; each has advantages and disadvantages and must be matched to the application.

Where the motion deliberately has not to provide constant velocity or positional adherence to the input form, intermittent drives may be used. These include the quick return, Geneva, intermittent gear drives, and the various forms of clockwork escapement mechanisms. Many couplings used make use of elasticity to allow for misalignment. In these instances it is unlikely that they will be of use if information fidelity or smooth unmodulated drive is needed through the link.

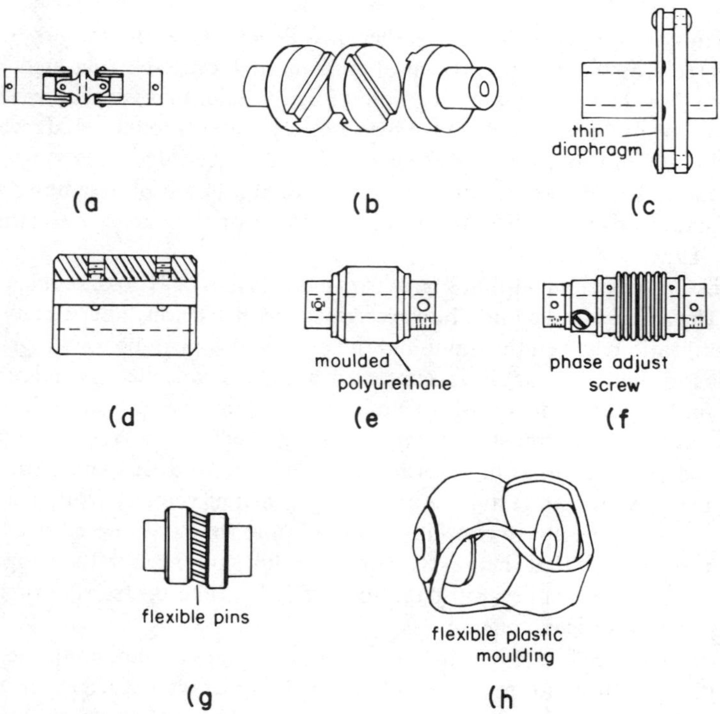

Figure 4.7 Selection of couplings used in instruments: (a) Double Hooke joint; (b) Oldham; (c) Membrane; (d) Rigid; (e) Molded flexible; (f) Bellows with phase adjust; (g) Flex-thane; (h) Kudriavetz

Gears

Although the toothed gear was in use in ancient times, it was not until more recently that it could be used extensively due to the relatively low cost resulting from mass production and good quality control of the form and size of the two parts involved.

Gears are used to alter (or maintain) the ratio of speeds between two turning or partly rotating shafts. Many forms enable the directions of the two shafts to be placed at any angle and displacement required. The main limitation of gears is that the shafts involved must be reasonably close together; this can be a serious shortcoming. Idler gears and unity ratio stages can be used to couple stages that are not close together. A combination of gears is called a gear train; it is housed in a gear-box.

The simplest form of gearing uses two smooth disks running in contact driving by friction force of the contact. Friction gears are necessary in many applications, for example, for converting the linear motion of running strip into rotation to drive an encoder, as parts of various mechanical computing devices, to subdivide a circular table, to drive paper through a chart recorder, or to drive a record turntable or tape recorder capstan — see also Sydenham (1968), and Jain and Sydenham (1980). Trylinski (1971) includes a chapter on friction gearing. Experience of Sydenham and Brandwijk (1966) indicates that friction gearing (Figure 4.8) can hold positional division of a 16:1 ratio step down to the order of one part in 20 000. The

Figure 4.8 Friction gearing used to subdivide a circle in the automatic manufacture of radial metrology gratings.

variation in ratio is apparently random in nature and absolute gearing synchronism cannot be maintained over a large number of revolutions. Best results are produced when the drive wheel has a rounded profile running on a flat surface, when the pressure is as light as possible to provide enough driving torque, and when the two materials are very hard, such as steel. High friction coefficient drive surfaces, like rubber, give poor constancy of gear ratio.

Where absolute synchronism or high torques are needed, the use of toothed gears becomes necessary. Spur gears are made in numerous forms and are generally familiar.

In instrument work the design of gears closely follows general practice with a few exceptions. One is the need to obtain precision operation. Gear trains in instruments and precision, small-size, closed-loop servo units will possess a certain degree of clearance between the teeth. This can introduce error of measurement or instability of a servo. It can be reduced by the use of antibacklash gears. These use two gears running adjacently that are spring loaded to cause the two halves to clamp around the mating teeth of the other gear. Whereas antibacklash gears do eliminate backlash, this form of device can introduce a degree of elasticity through the train, possibly resulting in a slack servo-loop. The backlash spring must hold the gears in contact against the friction of the train. It is often better to use the high grade gears for which backlash can be reduced by the closer tolerances of manufacture. As was mentioned in Chapter 3, the inertia of rotating parts may be significant; it must be considered when designing gear trains for dynamic use. Inertia rises as the fourth power of the gear radius and each stage adds inertia as the reflected inertia seen looking into it, multiplied by the gearing ratio. Multiple stage step-up gears soon accumulate significant inertia. The inverse obviously applies; step-down gear boxes see little inertia. High-ratio step-down boxes, however, will have increased backlash and the wind-up of the train becomes more significant as the elasticity of the gears, shafts, and bearings combine to form a more compliant through-link. As a general rule, the best response servos and the lightest action manual instruments result from the use of low gear ratios. The choice of more stages of lower ratio to obtain a given overall ratio may be better than to use a single large ratio step. Friction of step-down trains is generally of little consequence, but step-up units may load the drive excessively because of the multiplier effect of the train ratio. A gear train is also only as good as the bearings and box structure allow it to be. Considering that high quality gearing runs with around 5 μm runouts of concentricity and with pitch errors of similiar magnitude, it can be seen that the bearings and machining must be of the highest quality and the box be very rigid.

Gears are graded according to many codes, examples being the AGMA (old and new — American Gear Manufacturers Association) Admiralty, BSS, Mil Specs, IOS, and more. Specifying odd gears can be expensive. It is better to choose from standard ranges of diametral pitch and quality classes wherever possible. Reliance Gear (1972), Trylinski (1971), Mabie and Ocvirk (1958), Beggs (1955), Merritt (1975), and Michalec (1966) each cover the theoretical and practical design of the various forms of gears.

The designer can usually make use of marketed products that have been properly designed to obtain efficient and smooth operation. Manufacturers provide tables and other information to assist the designer to call up the correct centers, clearances, the lubrication to use and how to design gear trains with specified backlash. Reliance Gear (1972) contains a practical step-by-step guide to the design of fine-pitch gear trains, suggesting the following hints be followed.

The accuracy of gears can be gradually relaxed as the sensitivity to error reduces. The slowest speed gears are the most sensitive. Use spur gears where possible to make the most of the production precision that is available. Keep the number of meshes to a minimum to preserve transmission fidelity. Preload bearings to reduce backlash caused by bearing slop — which is not eliminated with antibacklash gears. Do not use pinned hub gears for high accuracy work. Sometimes gears can be matched by correct phasing to reduce the overall eccentricity error.

The finest pitch gears, theoretically at least must produce the most precise gear train but practice limits their production. As a guide, avoid teeth finer than 120DP or the finest above this that is consistent with the load requirements. Avoid mixing pitches in a train. Cleanliness of assembly and in use are vital.

High ratios are obtained in relatively small volumes using worm and wheel gears. These, however, may be inappropriate where the gear train has to transmit motion in both directions, as can occur in some servo drives. They can also be damaged easily in instruments where there are shafts that can be unwittingly forced by the operator or service man. Where this might occur, a friction clutch can be used to allow excessive torque to be dissipated without harm.

Gears are not restricted to use on full-circle wheels only. Part gears can be used where the motion needed is only part of a revolution or where intermittent motion is needed. The Leibnitz carry mechanism used in calculators and odometers is an example. Where precision length measurement or positioning is needed using gears, a rack and pinion is generally adopted. The pinion must run with little backlash and the guides be carefully manufactured to keep the system in alignment. The pieces of rack must by phased properly when butted together to extend the length range. Precision racks can provide relative readout or positioning to within one part in 100 000 (or a little better) uncertainty.

Cams

This is a two-part fundamental mechanism in which one part, the cam, imparts motion on a second element, the follower, that stays in contact with the surface of the cam. Cams take the form of shaped disks or plates turning on shafts, as curved slots cut around a shaft, as eccentrics within a yoke; they can also be formed as linear alternatives in which a follower rises and falls (Figure 4.9) in response to the shape of a linear plate profile moving relative to it. They are used to provide rotary to rotary or to linear motional conversion or linear to linear action. Cams are characterized by the feature that the output motion is only generated over a small relative magnitude. Cams cannot provide large strokes or rotations unless made equal or larger in size as the amplitudes needed. They find use in timing, function generation, for compensation, and as a means to produce motion from a source.

Fundamentally, cams are continuous surface forms of linkage. At any instance of position, the point of contact between the two parts and the support forms a linkage. The design of cams is well developed and they can be synthesized as well as analyzed using graphical and computational methods. Mabie and Ocvirk (1958) give a good coverage of cam design; Beggs (1955) and Faires (1955) provide less extensive accounts. Rothbart (1956) is a text devoted to cam design. Surprisingly, because cams are used extensively in instruments, Trylinski (1971) does not cover the subject.

Cam design basically attempts to realize the correct conversion action from the two components, but it is complicated by the fact that high-speed cams will impart significant velocity, acceleration, and jerk to the follower.

Figure 4.9 Linear corrector plate used to provide automatic compensation for calibration factor, in-scale non-linearities and possibly thermal errors in a wire length measuring device (Sydenham, 1969).

This can cause the follower to leave the cam surface and produce excessive wear. In other words, the correct functioning of the cam and its follower is lost unless the shape is appropriate for the dynamics involved. Various cam shapes provide different velocity, acceleration, and jerk characteristics. Typical forms used are parabolic, modified parabolic, and simple harmonic for relatively slow actions. For higher speed applications where jerk becomes significant, the forms used are cycloidal, half-cycloidal, harmonic, and eight-power polynomial. Each offers certain desirable features of operational acceleration and jerk magnitudes; see Mabie and Ocvirk (1958) for more detail.

Cams may run directly against the follower being in constantly sliding contact. This situation can be designed for by the use of hardened steels and other suitable materials. Alternately, a roller follower may be used. The follower can be a hinged rotating or sliding translating component. The cam, in disk or plate form, can be shaped to form a function required for applications of compensation or function generation. In these applications, velocity, acceleration, and the higher derivatives generally give no problems. Cams are also used to generate a shape that is 'followed' by optical means. This is used in furnace and other temperature controllers where the temperature must change with time in some specified manner.

Clearly, rapidly changing functions are not readily accommodated by cams because the follower has a finite size and will introduce, by virtue of its aperture width, some degree of integration. Cams are often required to rotate only in one direction. A reversed motion can become locked when the follower comes against a rise rate to which it cannot respond. Cam-shaped gear systems also exist to generate specific functions, but these are strictly gears, not cams.

Cams are also made in which the follower responds to the resultant of two inputs. These are referred to as 3-D or space cams. These need a spherical ended follower that is allowed to move in two translations as it is directed by the cam. Space cams are generally of complex 3-D shape and can be made by sintered forming methods or be cut by numerically controlled machine tools. These cams are unable to impart much force into the follower so they are often used with a servo that slaves to the follower's motion.

The dynamic behavior of cams can be complex because of the high value accelerations that can be present. These can cause the cam to appear elastic and can induce resonance in the cam body and in the follower and its spring. Beggs (1955) reviews this problem.

Flexible Connecting Drives

This classification is given by Trylinski (1971) to that group of motional transference mechanisms that use flexible elements to transmit motion. They may do this with or without conversion from rotary to linear regimes.

Some links use elasticity to provide the freedom of bending; others use kinematic, pivoted principles such as are found in a link chain.

Elastic types will not fail due to the bending stresses being excessive in their outermost 'fibres' if they have slenderness ratios of at least 300:1 (ratio of radius of bending to the thickness of the flexure in the direction of bending). This figure is only a guide. A calculation should be made of the stresses in the outer layer when the member is bent over the smallest radius in order to see if the material is not near to the yield point. There may be a jockey wheel added to keep the system taut or a guide pulley; often these are made too small in diameter. The stress level is as important as designing for adequate drive forces because excessive bending stresses will eventually cause the drive material to fatigue and break.

There are many alternatives available, each having particular advantages and disadvantages. Flat belts made of a material having adequate tensile strength and friction coefficient can be used with flat pulleys (actually the pulleys should be slightly curved, rising in the center, to cause the belt to track into the middle). More usually, instruments would use small section vee-belts or continuous round section belts made of reinforced plastics. O-rings make excellent belts; as they are made to tight tolerances, they provide smooth drives. If the forces to be transmitted are high, the pulleys and belts should be designed by calculation to decide the ability of the drive to operate with a realistic tension that will transmit the force without breaking. Round and vee-profile pulleys provide increased friction drive force without needing high belt tensions. Their design is laid down in texts on general machine design and is well defined. Trylinski (1971) is a satisfactory

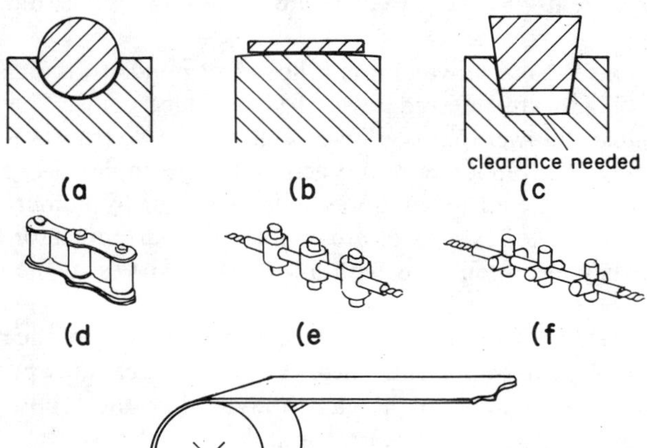

Figure 4.10 Cross sections of flexible connecting drives used in instruments: (a) Round, with or without reinforcement; (b) Flat (on slightly domed pulley); (c) Vee; (d) Metal link roller chain; (e) Polyurethane moulded — pinned belt; (f) As for (e), 3D pins; (g) Metal strip

source on flexible instrument drives. Figure 4.10 gives a range of drives used in instruments.

The friction drive belt depends upon the normal force exerted between the belt and the pulley surface in order to develop enough grip. These forces can sometimes be prohibitive and a better alternative may be needed. Furthermore, the drive may need to maintain a strict synchronized positional relationship between the two pulleys. For these cases friction belt drives are not satisfactory.

Synchronized flexible drive can be obtained by the use of toothed flexible belts running on appropriately toothed wheels. These are generally available only for comparatively large-size drives, the smallest being about 10 mm wide. Metal roller link chains also can be used but these, too, are not generally made for the size drives needed in the smaller instruments. Smaller special chains and toothed belts are made but their availability is not widespread. The user will have to search hard for a supplier (see Reliance Gear, 1972, and the British Machine Chain Co, Malden). Examples are the miniature oval-link chains that use specially shaped drive wheels in which the chain links lies keeping its crossed link shape, and the hook-link chains that use sprockets whose teeth pass through the links to provide drive. Fine metal chains can give backlash-free coupling at the force level needed.

Typical working forces for small chains are 1N, with links of 3 mm pitch. Small, plastic, positive-drive belts are also manufactured. One system uses stainless steel or cord-reinforced polyurethane molded belts, which comprise a round belt having bars formed across it protruding to each side. Another form has sides like a ladder shape. The bars nest into a single (or double) row of teeth on the drive pulleys. One dacron-reinforced belt will take 1 N forces, the unreinforced belt only 0.05 N. Some designs will carry a 15 N tensile load. There is also a toothed plastic belt that will drive sprockets set at 90° to each other without the need for belt twist. For illustrations of the above see Figure 4.10.

When only a small distance of angular motion is required, the belt need not be endless; further, it can be fastened to the two pulleys to obtain synchronized drive. Examples are the drive for a dial of a radio in which a cord connects the tuning control with the translating station indicator. Thin steel strips can also be used to couple wheels. As these can only drive in tension, a continuous strip or a biasing spring are needed. Trylinski (1971) gives several illustrations of the possibilities.

Flexible drives can offer extended distances between drive centers and the ability to provide gearing ratios; they can absorb vibration, attenuating it through the link (but they may lack stiffness if this is so); they can allow for misalignment between shafts, can be designed to allow slippage if the load is too great (or the converse), and can be designed to give a continuously variable gearing ratio by the use of tapered pulleys and changing radius pulleys. They are virtually backlash-free and will have relatively low inertia compared with gears trains because they have less mass in motion. They can, however, introduce vibrations from the unsupported belt, which can modulate the drive throughput of force or information.

Electromechanical Gearing

The present low cost of digital electronics and the general availability of very reliable and inexpensive optical and other shaft encoders has given the designer the ability to make more use of electromechanical gearing and shaft motion transference.

Combination of a shaft digitizer (Figure 4.11) that generates electrical positional signals as the shaft rotates and a motor that steps in synchronism with the generated signals forms the basis of an electromechanical gear box.

Illustrating the method with the use of the optical radial gratings, the disc attached to the input shaft can conveniently generate around 1000 to 10 000 pulses per turn at rates rising to at least 10^6 pulses/s; that is, shaft speeds varying from zero to 1000 rev/s for low resolution or to 10 rev/s for high resolution can be accommodated. The digital signals are then passed through a digital divider arrangement to form the required divided pulse rate to suit the speed needed from a stepping motor. Dividers are available as integrated circuits in which preset switches or electronic signals can be used to set any division number needed. Two IC's can give ratios from unity to 255.

The advantages of this alternative form of 'gear box' are that the input and output shafts need not be in close proximity or in any particular orientation, the power levels of the two sides of the link can be at any level needed, more than one shaft signal can be combined or driven, and the division ratio is very easily changed under computer control if need be. The ratio can be varied smoothly from one extreme to the other. Accuracy can be at any level desired and backlash is not present.

This form of system, however, can be prone to electrical interference, but it is quite possible, with circuitry effort, to make the error rate as small as is desired by the used of advanced communication transmission methods.

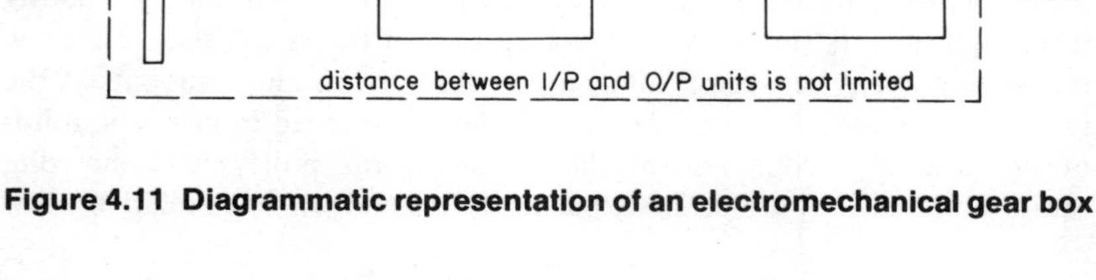

Figure 4.11 Diagrammatic representation of an electromechanical gear box

There is little doubt that this form of mechanical gear box will become more dominant in the future. The cost of the electronic parts will be reasonably small and the mechanical parts needed are few — a shaft carrying an encoder disk to couple to the output of a mechanism and an output shaft of the stepping motor. It is to be expected that such gear boxes will be cheaper than conventional units, especially where high precision is needed. These will, of course, need a source of electrical power, which may exclude them from many applications.

References

Beggs, J. S. 1955. *Mechanism*, McGraw-Hill, New York.

Braddick, H. J. J. 1954. *Mechanical design of laboratory apparatus*, Chapman and Hall, London (1st edition – 1954).

Burstall, A. F. 1970. *A history of mechanical engineering*, Faber and Faber, London.

Doughtie, V. L. 1966. *Elements of mechanism*, Wiley, London.

Drysdale, C. V. and Jolley, A. C. 1924. *Electrical measuring instruments — Part 1 Commercial and indicating instruments*, Ernest Benn, London.

Faires, V. M. 1955. *Design of machine elements*, Macmillan Co, New York (1st Edition 1934).

Ferguson, E. S. 1962. 'Kinematics of mechanisms from the time of Watt', *United States National Bulletin 228*, Paper 27, p 185 – 230. Smithsonian Institution, Washington, D.C.

Fremery, J. K. and Doden, K. 1978. 'Active permanent magnetic suspensions for scientific instruments', *J Phys E: Sci Instrum*, **11** (2), 106 – 113.

Furse, J. E. 1981, *"Kinematic design of fine mechanisms in instruments"*, J. Phys. E.: Sci. Instrum., **14,** 264 – 272.

Geary, P. J. 1955. *Knife edge bearings*, Sira Institute, Chislehurst.

Geary, P. J. 1961. *Instrument ball bearings*, Sira Institute, Chislehurst.

Geary, P. J. 1962. *Fluid film bearings*, Sira Institute, Chislehurst.

Geary, P. J. 1964. *Magnetic and electric suspensions*, Sira Institute, Chislehurst.

Harris, T. A. 1966. *Roller bearings analysis*, Wiley, New York.

Jain, P. K. and Sydenham, P. H. 1980. *"Radial metrological patern generating engine"*, J. Phys. E.: Sci. Instrum., **13,** 461 – 6.

Jeffery, G. J. and Sydenham, P. H. 1973. 'Stability of strain-meter mounts', *Geophys J R Astr Soc*, **32** (2), 185 – 193.

Lent, D. 1970. *Analysis and design of mechanisms*, Prentice-Hall, Englewood Cliffs.

Mabie, H. H. and Ocvirk, F. W. 1958. *Mechanisms and dynamics of machinery*, Wiley, New York (3rd Edition – 1975).

Merrit, H. E. 1975. *Gear engineering*, Wiley, New York.

Michalec, C. W. 1966. *Precision gearing: theory and practice*, Wiley, New York.

Neubert, H. K. P. 1975. *Instrument transducers — an introduction to their performance and design*, Clarendon Press, Oxford.

Peatfield, A. E. 1951. *Engineering components and materials*, EUP, London.

Pollard, A. F. C. 1929. *The kinematical design of couplings in instrument mechanisms*, Adam Hilger, London.

Reliance Gear. 1972. *Catalogue RG5*, Reliance Gear Co Ltd, Huddersfield. (Also similar catalogues by PIC Design Corp, New York).

Rothbart, H. A. 1956. *Cams: design, dynamics and accuracy*, Wiley, New York.

Shields, J. 1976. *Adhesive Handbook*, Newnes-Butterworth, London.

Sydenham, P. H. 1968. 'Linear and angular transducers for positional control in the decametre range', *Proc IEE*, **115,** 1056–1066.

Sydenham, P. H. 1969. *Manual and automatic large-scale dimensional metrology*, PhD thesis, University of Warwick.

Sydenham, P. H. 1974. '2000hr comparison of 10 m quartz-tube and quartz-catenary tidal strainmeters', *Geophys J R astr Soc*, **38,** 377–387.

Sydenham, P. H. 1979. *Measuring instruments — tools of knowledge and control*, Peter Peregrinus, London.

Sydenham, P. H. and Brandwijk, J. F. 1966. 'Production of radial line gratings using a mechanical photographic apparatus', *J Sci Instrum*, **43,** 380–382.

Timoshenko, S. and MacCullough, G. H. 1959. *Elements of strength of materials*, Van Nostrand, New Jersey. (Numerous editions and printings exist since the 1st in 1935).

Trylinski, W. 1971. *Fine mechanisms and precision instruments — principles of design*, Pergamon Press, Oxford, translated by A. Voellnagel from Polish original.

Turner, G. L'E. 1973. *Van Marum's scientific instruments in Teyler's Museum — Descriptive catalogue*, Noordhoff Int Pub, Leyden.

Zimmerman, J. R. 1962. *Elementary kinematics of mechanism*, Wiley, London.

Chapter 5
Putting elasticity to use

The place of flexure elements in instruments

In the previous chapter mechanisms used to assemble instruments were divided into two groups, namely, kinematic and elastic. Those based on kinematic principles were discussed. In this chapter it is the other group that is of concern: those that deliberately make use of elasticity of components to achieve a given function. Flexing action produced by kinematic mechanism, such as a roller chain, is covered in the previous chapter.

By appropriate choice of compliance, component shape, and method of use, spring elements (often called flexures) can also provide many of the functions described in the previous chapter. Kinematic components provide arrangements in which all components parts are assumed to be rigid without exhibiting any noticeable degree of elastic give; elastic design deliberately makes use of elastic properties to allow movement where needed.

In many instances a flexure method is to be preferred over a kinematic arrangement; this generally occurs when only limited movement between two parts is needed. For example, a crossed-strip hinge provides a limited angular range hinging action, having negligible backlash and frictional force. It does not need lubrication and is not affected by dust or dirt. It is very simple to make and often does not require tight manufacturing tolerance.

Flexure elements can be devised in numerous shapes and forms. Good design is always based on proper understanding of the compliance of a flexure element in its various directions of bending and compliance values of associated other elements. In some cases it needs to have a high ratio of compliance in one direction with respect to another; in other cases it needs to be the same. The addition of later items, such as protective covers, additional small loads, and further forces from sources other than those designed for, could considerably alter the final performance of an elastic design.

Flexures elements appear to be deceptively simple. They are easy to devise and install and they will often provide quite adequate performance with only the barest of theoretical understanding being applied. However, a thorough appreciation of the factors involved will assist the user to understand their limitations; they are not perfect elements.

The necessary relevant basic theory pertains to that of the deflection and stresses of beams and plates — commonly called strength of material or beam theory. To this must be added knowledge of the imperfections of real materials because these introduce such effects as damping, friction, hysteresis, fatigue, fracture and nonlinear deflection versus load characteristics, subjects not normally covered in beam theory texts. Fundamental material (which is of civil and mechanical engineering structures design) is to be found in such works as Timoshenko and MacCullough (1959), Timoshenko and Gere (1961), and Roark and Young (1975). Faires (1955) provides a shorter but basically adequate account of the strength of materials theory involved. Such texts, being written for structures designers and not instrument applications, generally place greater emphasis on stress than on the deflections so important to instrument design.

The design of springs elements, which commonly concentrates upon helical and flat power springs, as opposed to instrument springs, is dealt with by Chironis (1961) and Wahl (1963). Spring suspensions and flexure devices, for which the emphasis is upon fine mechanism, are covered in Eastman (1937), Geary (1954), Geary (1960), Andreeva (1966), Trylinski (1971), and Sydenham (1984).

Geary (1964), Frazier *et al.* (1974), and Fremery and Doden (1978) cover another type of spring suspension: that provided by magnetic and electric fields. Dorien-Brown (1971) is one review of air springs.

Many other works include mention of flexure arrangements in sections ranging from chapter length down to a few paragraphs. These include Rolt (1929), Whitehead (1934), Thorpe (1953), Braddick (1954), Beggs (1955), Weinstein (1965), Jones (1967), Jones (1968), and Neubert (1975). Hugill (1978) compares several spring-mass systems used in gravimeter designs. Equations of sensitivity are given in each case. Gast (1974) includes discussion on flexures. Dratler and Block (1972) is an interesting report.

Flexures design studies are usually based upon the assumption that the force-deflection characteristics of the materials used are linear. Such a basis, can, indeed, give the designer a very wide choice of workable ideas, but on occasion it might be beneficial to make use of nonlinear materials and processes. Eisley (1963) provides an entry point into the literature of this area of knowledge, giving about 90 references to published papers. Schmidt and Da Deppo (1973) may also be useful in this area.

Additional references are introduced in their context as specific applications arise in the following discussion of this chapter.

There is, therefore, a considerable amount of published design information available but, as can be seen, it includes very little of recent times. This should not be taken to imply that the information has become outdated in any way. New interest in the rigorous design of electromechanical instruments has been a strong reason for renewed attention. There is little likelihood of a basically new flexure element being invented, most having been devised many decades ago. It is intended that this chapter will provide a comprehensive overview of the many possibilities for elastic design. Combination of several flexure mechanisms leads to more options, many of which have yet to be applied.

The apparent simplicity of flexures can hide the fact that they are not perfect elements. As an example, a parallel strip support for a translating table can easily support a mass of many kilograms, providing translational sensitivity and stability of nanometer order without backlash being apparent. However, temperature effects may alter the inbuilt spring-rate quite considerably, causing errors in other ways; choice of the wrong material can introduce significant hysteretic backlash.

As flexure systems deliberately introduce spring compliance into a system where mass is involved, it is obvious that linear second-order and higher dynamic systems may be produced. The natural frequency of such systems becomes an important parameter where the system has to follow dynamic inputs in a faithful manner (Sydenham, 1982, 1983). Damping may be deliberately introduced to prevent reasonances building in amplitude or to make the dynamic response at least critical so as to exhibit response characteristics of first-order nature. It is feasible by appropriate design and ingenuity to alter the natural frequency without altering the compliance. For the same reasons it may be necessary to keep the supported mass to a minimum so that the natural frequency is kept as high as possible.

Design parameters of the flexure element

Before considering practical flexures and their application, it is appropriate to discuss several basic design parameters of flexure elements since these apply to all shapes and forms of system built with them.

Flexures are formed by the use of a mechanical member (or possibly other spring form) in which design deliberately allows the member to deflect in specific directions. It is usually imperative that deflection occur only within the elastic region of the material's deformation (Figure 5.1). Although texts on materials give the general impression that the Hookean elasticity law applies in practice, there nevertheless is always a small amount of hysteresis; a strained member does not come back to exactly the same strain condition for a given stress excursion. Special materials, those with a reduced mechanical hysteresis (see Chapter 2), exhibit a lower level of this effect. In many applications of flexures in instruments, however, extreme reproducibility is not important. Tovey (1974) discusses the materials used to form flexure elements.

It is therefore important that the flexure members cannot be deflected beyond the chosen allowable elastic limit. As a rule the smaller the deflection, the smaller the residual strain error. Plastic deformation, if allowed to occur, will introduce permanent set with subsequent zero shift errors arising. Design, for this reason, often calls for stops and other devices to be incorporated to prevent over-deflection. Assembly and transport can often give rise to the greatest deflections that might be encountered. As a rough rule of thumb guide, a bending flexure elements should not be deflected into a radius of curvature smaller than around 300 times its thickness in the direction of bending. Beyond this level the outer fibers will be stressed beyond their elastic limit, causing surface cracks to begin. If this is important, the bending moment caused by expected flexing levels should be calculated for the material used.

As a flexure deflects it will provide a force reacting with the direction of motion. This is the flexure system compliance. It is produced fundamentally by the elasticity of the material, the effect of which is modified by the total flexure system arrangement. As will be mentioned below, it is possible to devise flexures that provide positive, negative, and even zero forces as well as nonlinear displacement characteristics.

Using strips in tension will increase the stiffness according to the load (Eastman, 1937). Conversely, they can be used in longitudinal compression quite satisfactorily provided Euler buckling does not occur (see previous chapter). The basic elements still undergo positive or negative deflections

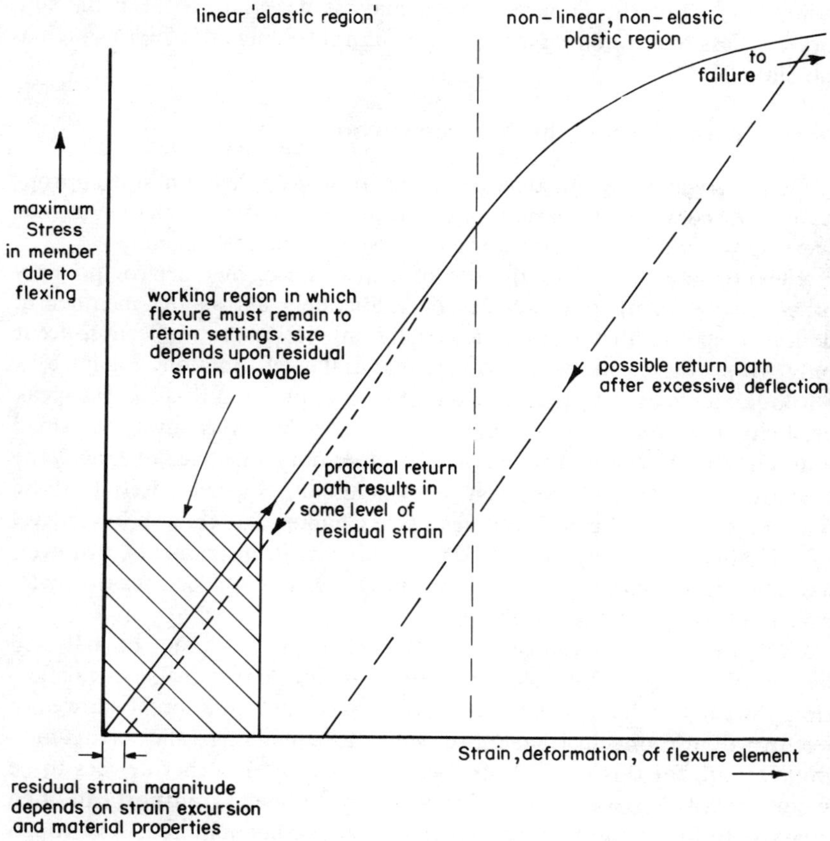

Figure 5.1 Stress-strain relationships for materials used to make flexure elements

toward or away from the elastic limit, but the net effect of the whole flexure system is one of these options.

The fundamental parameters of a flexure material are its elastic modulus, shear modulus, hysteresis level, elastic limit endurance limit to constant flexing (especially when in conditions of simultaneous corrosion effects), thermal coefficient of elastic modulus, and such other parameters in any other energy domain as were listed in the material properties checklist given in Chapter 2. Rolling friction is sometimes a parameter for which more information is published in Minato *et al* (1969). Gohn and Fox (1965) discuss stress relaxation in springs. Carson (1937) is about measurement of elastic drift in springs.

When very small displacements are needed, a method used by Jones (1967) might be applicable. It rests upon the principle that an elastic material that is squeezed by a clamp will exhibit a small length change. Poisson's ratio links this action. This method gives fine control to clamping screws on threaded adjustments such as micrometer heads The same principle was employed by Baird (1968) to temperature-compensate an Invar rod with a steel wire cover soldered to its outer surface.

When flexures are made from thin hardened and tempered strips, they can tear easily at connection points. Careful edge finishing is needed to reduce this failure mode. The material must also be cut to suit any directional grain that exists.

Materials used for flexures are generally the metal carbon spring and stainless steels and the copper alloys and the nickel-iron alloys. In some applications fused-silica (fused-quartz is usually synonymous) is used as the flexure element. In less exacting applications, plastic, rubber, paper, and fiber elements might be suitable.

The above parameters are concerned with the properties of the material used to form a flexure. Having selected the material, it is then necessary to consider the properties of the flexure system as a whole. Some factors that may be relevant include the force-deflection characteristic, the range of motion, required parallelity of pseudo-sliding action, stability of a center of rotation with angular deflection, and effect of load on the system properties because the system may stiffen up with increasing load being supported. Another important design parameter is to be sure that the system can be manufactured to the tolerances needed and that it can be assembled without damaging the flexure element in some way. Preform in flexure elements can be bothersome at times. Elements may need to be heat-treated to stabilize them before use and be assembled using jigs to prevent overstressing.

Design of a flexure system, like that of a kinematic support system, should always consider the sources of error that might arise in normal operation.

At present, and perhaps it will always be so, the design of a flexure system is approached initially as an art based upon experience with the possibilities open to the designer. A possible arrangement is first selected that may fulfill the need. This is then analyzed as a science to establish the sensitivity, load-carrying capacity, limits of required action, and so on. If the design is not quite suitable, it is then modified. It would be an obvious advantage if computer-aided design programs were available for use in optimizing standard design configurations (of which there are several kinds) in the direction that a designer requires.

Design equations for most types have been reasonably well developed, but the designer usually must spend considerable time in locating the appropriate study. Many reports on flexure elements usage do not include a theoretical account, their design often being apparently raised by intuition and cut-and-try prototype building procedures. Of the references to general accounts given above, those including a well developed basis of theory to support a design are Chironis (1961), Eastman (1937), Andreeva (1966), Geary (1954), Geary (1960), Thorpe (1953), Trylinski (1971), Wahl (1963), and Weinstein (1965). Articles on specific designs are referenced later where they are discussed. Table 5.1 provides deflection and stress expressions for a few of the commonly met elements. Note that point load and distributed load cases each follow the same general laws.

For flexures to be theoretically designable, it is necessary that their stress levels be determinable for all parts of the system. In the kinematic design procedure, forces are arranged so that they act at known, well-defined points of interaction. In flexure design, deflection takes place over a distributed region. To assist design it is often expedient to structure the flexure elements of the system as members having dramatic changes of size at given positions so the compliance alters very significantly. It can then be assumed that the flexure member is deforming elastically with the end mounts being adequately rigid. From the viewpoint of fatigue strength, however, such sudden geometrical changes give rise to locations in the system where stresses can be raised. Thus, flexure designs needing high numbers of highly stressed cyclic movements will require a more sophisticated treatment to allow for graded compliance changes along the flexure element.

It is stated by some authors and users of flexures that the applications requiring the greatest dimensional stability should use flexures that have end mountings made in the same piece of material. (The usual method is to clamp the flexure end between rigid plates.) This assertion has not been adequately proven by rigorous experimental procedures and it must be left to the user to decide if this additional cost is necessary and justified.

A similar argument exists as to whether metal springs elements give better performance than fused-quartz springs. Experience with both has shown that poor reproducibility of zero with time is more likely to be a factor of overall design than of the material used. It is possible to find evidence that suggests that either material group is the better to use for precision instrument springs. As the problem reduces to one concerning the hysteresis and temperature properties of the materials, it can be seen that certain metal alloys (see Chapter 2) are far superior to fused-quartz in this respect.

A key parameter that tends to be overlooked in such comparisons is the level of flexure excursion. Small movements of a spring element such as occur in a tidal gravimeter and a tidal tiltmeter, will give rise to much smaller levels of hysteresis when, compared with the performance of an exploration gravity meter of a borehole tiltmeter, in both of which the flexure element must range over larger excursions and be more severely handled.

Table 5.1
Expressions for commonly met beams

Point loading	Deflection δ	Maximum Stress σ	Type
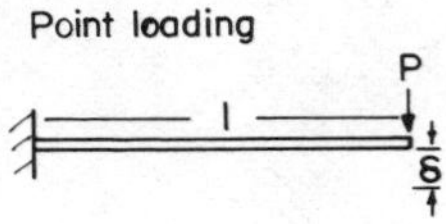	$Pl^3/3EI$	Plh/I	cantilever with point load
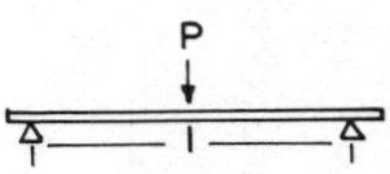	$Pl^3/48EI$	$Plh/4I$	simply supported beam with central point load
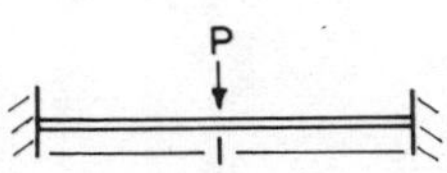	$Pl^3/192EI$	$Plh/8I$	point load applied to centre of beam with ends constrained rigidly
Uniformly distributed loading			
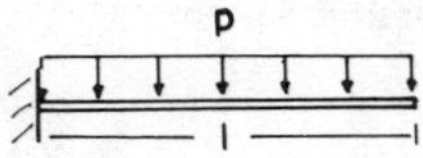	$pl^4/8EI$	$pl^2h/2I$	cantilever with uniform distributed load over whole length
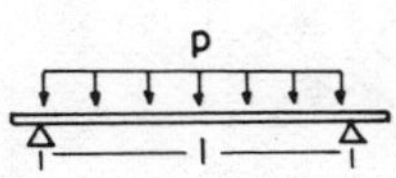	$5pl^4/384EI$	$pl^2h/8I$	simply supported beam with uniformly distributed load only between supports
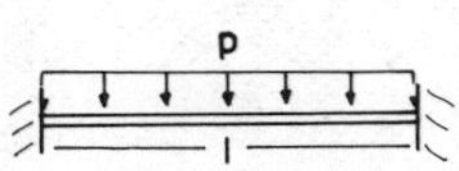	$pl^4/384EI$	$pl^2h/12I$	uniformly distributed load applied to whole beam with ends constrained rigidly

where I - second moment of cross-section (see Table 4·2)
E - Youngs modulus of beam material
h - maximum distance of any point in beam from neutral axis
P - point load
p - load per unit length

further expressions for other loadings, for torsion on beams and for diaphragms are variously available in :
Neubert (1975), Roark and Young (1975), Timoshenko and MacCullough (1959), Trylinski (1971).

Types of flexure element

A flexure can provide controlled movement in a system of flexure elements using one or more of three basic modes of flexure (Sydenham, 1984).

It can be deflected in pure transverse bending, in pure axial torsion, or it might be deformed elastically in its longitudinal compressive or tensional mode (Figure 5.2(a)). In practice, combinations of these three actions occur to a lesser or greater extent depending upon the flexure system arrangement. Too often designers (and quoted formulae for flexure arrangements) do not recognize this combinational feature. For example, a simple single flexure strip being used at the top end of a clock pendulum to allow it to swing in one preferred direction also supports the mass of the pendulum. The flexure element is, therefore, simultaneously stressed in both bending and tension. The tension component can considerably stiffen the rotational compliance in the bending direction.

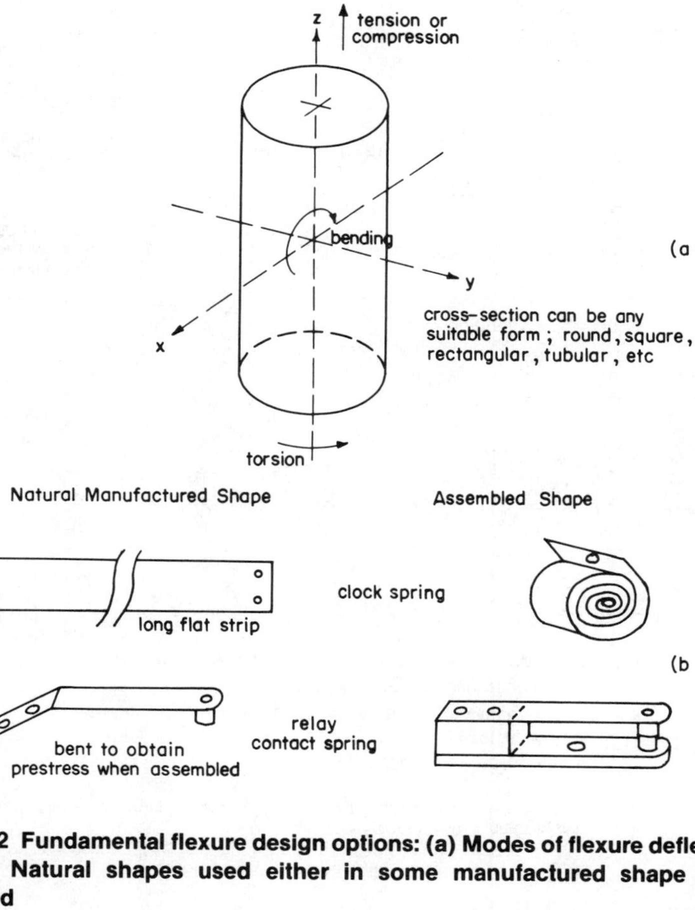

Figure 5.2 Fundamental flexure design options: (a) Modes of flexure deflection, (b) Natural shapes used either in some manufactured shape or preformed

The helical coil spring is a torsional arrangement in which torsional flexure is converted to provide linear spring-force action. Torsional flexure systems can make use of other than strip or other solid members. Two wires, placed apart, can produce a similiar action — these are called bifilar suspensions.

The third class, wherein longitudinal tension or compression used without transverse bending action of the flexure, is applicable when stiff flexures are needed. Simple examples would be the use of a bolt to hold two parts together yet allowing some additional elasticity to enable it to cope with excess tension loads; or a rubber isolation mount. Torsional systems often stress the individual members in this way. Such methods may dictate the use of materials with low values of Young's modulus so that they provide reasonable compliance from practical cross sections of material. Their use is, of course, dependent on the requirement and materials suitability.

With the basic elastic kind of action decided, the designer then has a further fundamental option available. Many flexure systems use the element in a stress free state when the system is at its rest position. For example, a helical spring may be coiled so that the turns just nest together so that with no extension the spring material is in a relatively stress-free condition (see Figure 5.2(b)). Elastic energy is then stored as the spring is tensioned. Another example is the use of clock spring to provide drive energy for a clock. In its unassembled state the natural shape of the flexure is a long flat strip. As it is wound up it takes on the curved shape seen in the assembled clock.

The alternative arrangement is to prestress the spring material into a natural shape that operates in the reverse manner to the obvious situation exemplified above. A torque motor (that needed to drive a mechanical clock) can be produced by precurving the strip so that its natural shape is a small radius, that is, the unassembled spring rests as a tight coil that is then unwound as it is deformed to store energy. Springs holding electrical contacts in relays are usually preset to some extent so that their assembled state holds the contact in the correct position applying a force to the other contact.

Thus, design of springs must also allow for this initial material condition of the flexure elements used, as well as the system arrangement in which they are used.

Yet another distinction can be drawn between the flexure system that is purposely intended to store energy in order to produce a force on demand and that which uses flexure action as a mechanical linkage. In the latter case it is often necessary to design for minimum stored energy as deflection occurs, storage and its resulting force-producing action being an unwanted parameter. Often the flexure system simultaneously provides both mechanical linkage and energy storage functions; for example, a cantilever flat strip spring provides return action and guidance.

Through the use of a suitable design of spring, it is possible to obtain mechanical magnification of small displacements. Such springs are generally of the type in which a small length change of a tensional spiral spring produces a rotational motion in the center, or at the end of the spring element. The rotation can then be used to operate a suitable length output arm or the mirror of an optical lever. Torsional magnifiers are discussed in Geary (1960).

With these general differences clarified, the discussion can now turn to descriptions of specific systems. In practice there are a very great number of possibilities, so it is not feasible to list every combination and arrangement that has been reported. Those presented are arrangements for which adequate published information exists for them to be used and designed using rigorous principles. It is possible to seek other sources not covered and perhaps develop the necessary theory. Collectively, the referenced literature presented here (when reviews included within several of the citations are also recognized) provides a very large range of diagrams and applications of flexure elements in the design of instruments. What constitutes an instrument is a matter of personal definition in this area, for flexure elements find as much use in power supporting applications as they do in fine mechanisms.

Helical and Spiral Springs

The design of spiral (helical, coil) springs is well covered in numerous commonly available sources because they form part of the training of engineers in the general design of machines. For example, each of Chironis (1961), Andreeva (1966), Faires (1955), Geary (1960), Trylinski (1971), and Wahl (1963) can be consulted for the theoretical design of the familiar cylindrical coil spring. Such springs are generally intended for use as pure tension or pure compression flexure elements (Figure 5.3(a), (b)) but they can also be used as torsional or bending elements.

The above references discuss the key spring design parameters to obtain a spring with a given spring rate (compliance, stiffness), adequate range of deflection, adequate fatigue life, and given maximum stress levels in the material, each applying for the range of materials normally available. The more general machine design texts do not usually consider such factors as effect of temperature on the spring rate (see later chapter on error sources in instruments) or the need to reduce the hysteresis effects — two factors that are vital to instrument design.

Figure 5.3 shows several of the many shapes and forms that such springs may take.

Helical torsion springs (Figure 5.3(c)) are formed into shapes that are similiar to the coil spring, but the forces are applied to the coil as a torsion rather than longitudinal extension or compression. In this use the basic mode of flexure of the spring material is not torsional, as it is in the tension helical spring, but occurs as bending action.

In this use the turns are required to rotate with respect to each other; practical design has to allow for any friction (the binding effect) that may occur when the coils slide against each other. Such springs are often made with spacing between turns. Another factor to be considered is that as the spring is twisted tighter in torsion, the effective radius reduces. Long torsional helical springs may buckle; guides are often employed to support the spring in which case the added effects of friction must not be forgotten in sensitive system design. Wahl (1963) suggests that buckling is avoided if the spring is not turned at the torsional input by more than 0.87 of a turn for length to diameter ratios that are short, the input rising to 1.25 turns for longer springs having the same diameter. This can only be taken as a guide; the references cited by Wahl should be consulted for more detail.

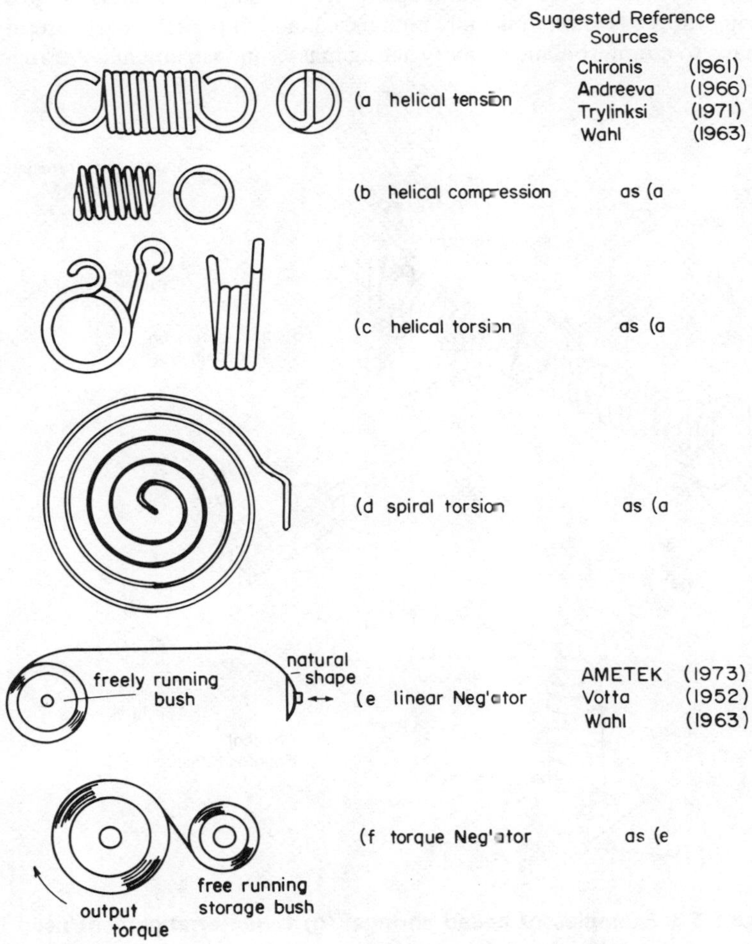

Figure 5.3 Examples of coiled springs (continued on next page)

Spiral springs are those wound springs in which flat material is used (Figure 5.3(d)) that is curved in a decreasing radius and where all turns lie in a flat plane. A clock or an electromechanical indicator restoring spring are in this group. These are generally referred to as hair springs when the turns do not come into contact — the spiral spring design, which exhibits the least damping action. Hair springs can also buckle at high deflections.

It is in this class that the Neg'ator spring design fits. This preformed spring was invented and reported as recently as 1952 by Votta (1952). It has the interesting and somewhat hard to visualize characteristic that it can provide linear tensioning action, exhibiting constant force with displacement (as opposed to most springs in which tension increases with displacement). By suitable choice of bush size of its support (Figure 5.3(e)), it is possible to tailor the design to provide very slight negative or positive spring rates based on unusually high base loads. It is particularly useful as a means to counterbalance gravity acting masses in instruments. Arranged as

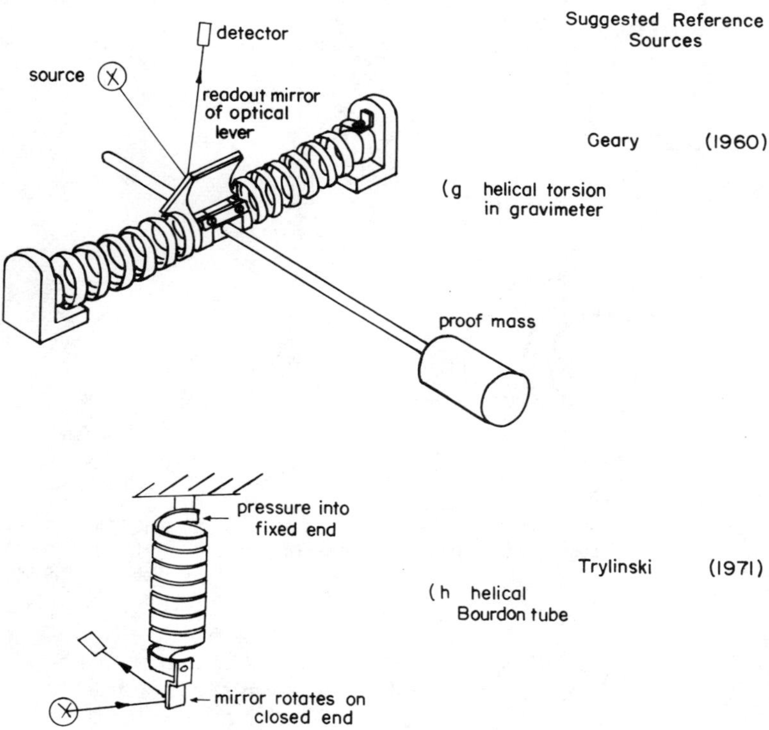

Figure 5.3 Examples of coiled springs: (g) Helex arrangement used in a gravity meter to suspend the proof mass. (h) Bourdon tube form used for ambient pressure measurement

a torque producing device (Figure 5.3(f)), instead of giving purely linear action it forms a constant-torque drive motor with a large range of output turns. In Britain it is marketed as the Tensator spring. Typical commercial design data is available in AMETEK (1973). The torque motor form of spring has been analyzed as a dynamic motor in Sydenham (1969) where it is shown that the commercially offered width to thickness ratios and suggested spool diameters lie far from the optimum values needed to obtain minimum rotational inertia. Given more optimum spring material thickness and width, the use of small diameter spools in such motors can provide very responsive torque sources; energy stored can be dangerous in use and protection is needed to prevent damage should a spring break or come away from its end constraints.

The design of spiral springs is covered in Andreeva (1966), Geary (1960), Trylinski (1971), and Wahl (1963). Each of these sources covers some of the same material, but they each additionally provide somewhat different design information. It is necessary to select the appropriate work for the task in hand, for none of them appears to give a coverage that caters for all likely needs; there are just too many spring options in existence. Trylinski (1971) concentrates on fine mechanism springs, that is, those including hair springs and main springs for such instruments as clocks and watches. Geary (1960) is useful for its general introduction and for the bibliography provided. It provides theory of torsional systems using bifilar and multifilar suspensions. Wahl (1963) and Andreeva (1966) give a wide general coverage of all types.

Several highly sensitive instruments make use of a spiral form of spring in which the turns are not in contact and lie in a helix. Figure 5.3(g) gives an arrangement used in one gravity meter to suspend the proof mass. Figure 5.3(h) is that of a somewhat similar use, but where the form of a spiral Bourdon tube is used for ambient pressure measurement. Reference sources are given in the figures.

Flat Strip Spring Systems

Flat strips used in bending and torsional modes of basic flexure, as with previously treated springs, can also be used as energy storage elements or as supportive mechanisms. In many applications they serve both purposes together.

With so many options reported, it appears most expedient in this brief introductory review to present schematic illustrations of a selection of options describing some of them in detail to bring out certain design features. Figure 5.4 provides various views of flat-strip springs used to provide energy and support. Several use the flat strip in a coiled fashion; it is not possible to entirely separate spring forms into unique groups.

The first type shown in Figure 5.4(a) is the use of a strip in which one end is secured so that rotation at the fastening is prevented. Deflection progressively occurs along the strip, increasing in amplitude as the force moves outward. The force-deflection characteristic depends upon the shape and thickness of the strip. It is nonlinear for a parallel, uniform strip. To linearize the spring rate, it can be tapered in thickness and width. Leaf springs used in vehicle suspensions are actually a very tapered cantilever spring that is cut into sections and stacked in a pile. Early expensive vehicles further tailored the characteristic by tapering the spring material in thickness as well as width. The design of these is straightforward. The strip can be regarded as a beam with one end (or two ends if needed) fastened rigidly. Another option for altering the rate characteristic is to let the step progressively lay onto a shaped support as it deforms (Beggs, 1955). Figure 5.4(b) shows the strip used in the same way as a beam with both ends supported. In this option the ends can be simply supported so that rotation is allowed at the ends. Deflections are not linear with load unless the beam is suitably proportioned in thickness by the means mentioned above.

When cantilever springs are used as small rotation locating mechanisms, the effective pivot point, called the instantaneous center of rotation, moves in a curved path. Cantilever flexures can also be used with the end of the beam being placed in torsion. A tensioned beam used this way with long length and narrow width is called a tautband suspension.

Figure 5.4(c) shows use of strip to provide bending while supporting a load. As has already been mentioned, the bending stiffness is increased as the load supported is increased. Many expressions given in texts do not allow for this. The pivot action provides an instantaneous center of rotation that moves in a circular path as does a cantilever support. Geary (1954) should be consulted for reference sources on this design. Trylinski (1971) and Eastman (1937) are also relevant.

Proof rings, which form the force to displacement conversion element of load cells, are formed of shaped rings, tubes, rods, and more, to which are attached a displacement sensor. The sensor might be a simple mechanical indicator or it could be an electrical output device. In either case the design aims to obtain linear or other specific conversion characteristics by suitable choice of mechanical force element. Figure 5.4(d) shows a circular proof ring formed from thick strip by machining it from the solid or by forming flat plate.

Displacement of a table in one axis while the other five degrees of freedom are substantially held in constraint can be obtained by use of a parallel strip suspension shown in Figure 5.4(e). These take several forms of which the simplest variation is shown. Study of that given, a four-strip suspension, will reveal that as the amplitude of displacement rises the table being supported does not follow a perfect linear translation but traces out a curved path in which it rotates and rises and falls. These effects reduce as the length of the strips increase with respect to the amplitude of movement needed.

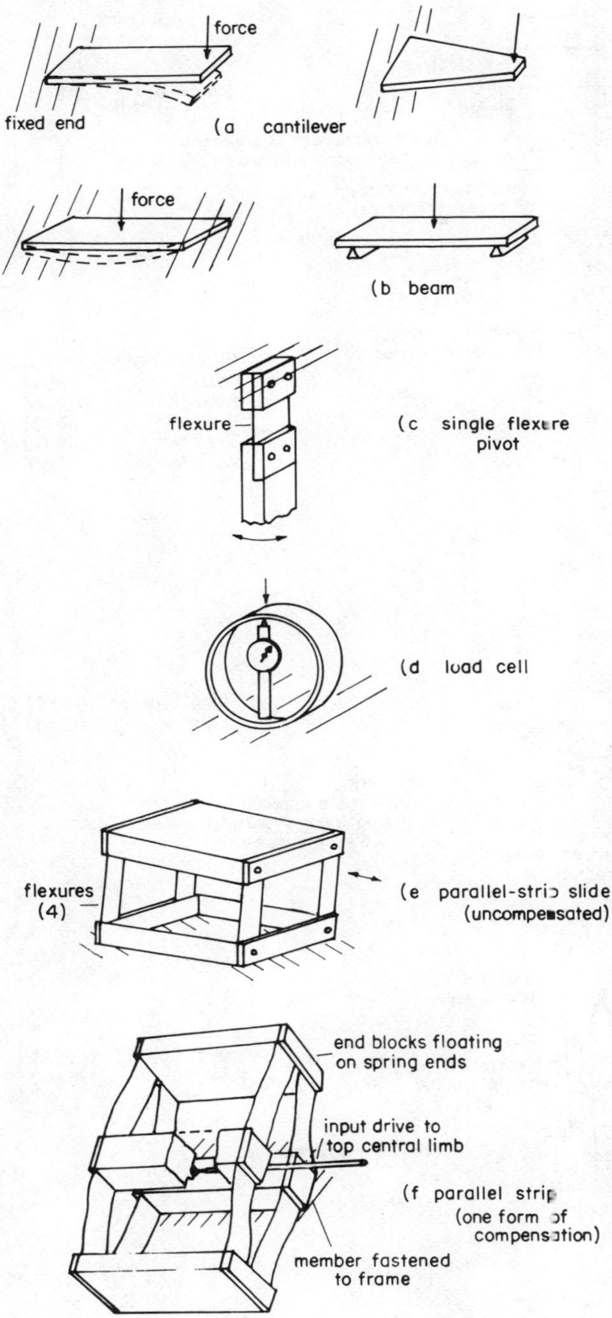

Suggested Reference Sources

Andreeva (1966)
Geary (1954)
Neubert (1975)
Trylinksi (1971)

Eastman (1937)
Endreeva (1966)
Geary (1954)
Trylinksi (1971)

as (a

Geary (1954)
Jones (1951)
Jones/Young (1956)
Neubert (1975)
Thorpe (1953)
Trylinski (1971)
Wahl (1963)
Weinstein (1965)

Geary (1954)
Jones (1951)
Jones/Young (1956)
Neubert (1975)
Trylinski (1971)

Figure 5.4 Examples of flat strip springs (continued on next page)

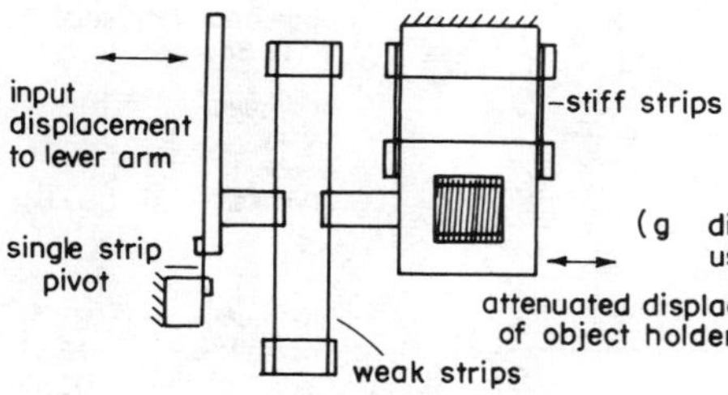

(g displacement attenuator using strip suspension

Suggested Reference Sources	
Sims et al	(1971)

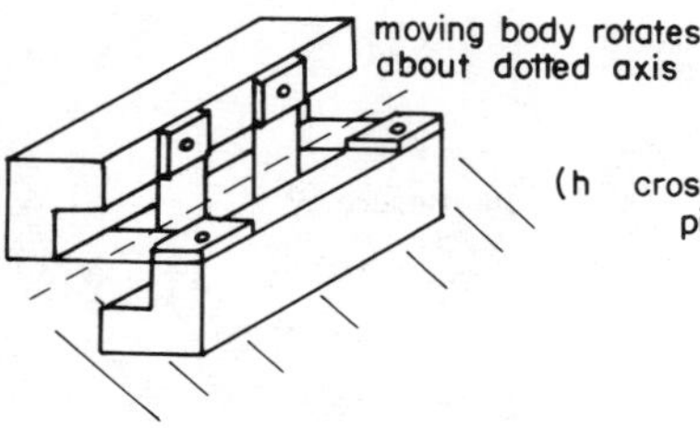

(h cross-strip pivot

Suggested Reference Sources	
BENDIX CORP.	(1975)
Geary	(1954)
Haringx	(1949)
Neubert	(1975)
Trylinski	(1971)
Weinstein	(1965)

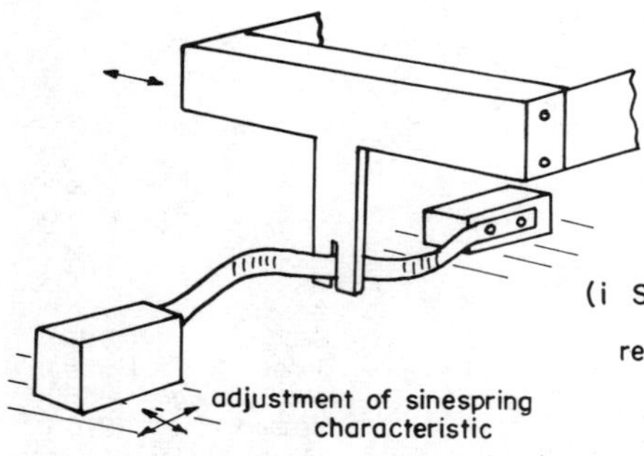

(i Sinespring (used here to reduce restoring force of parallel strip suspension)

Suggested Reference Sources	
Eijk / Dijksman	(1977)
Geary	(1954)

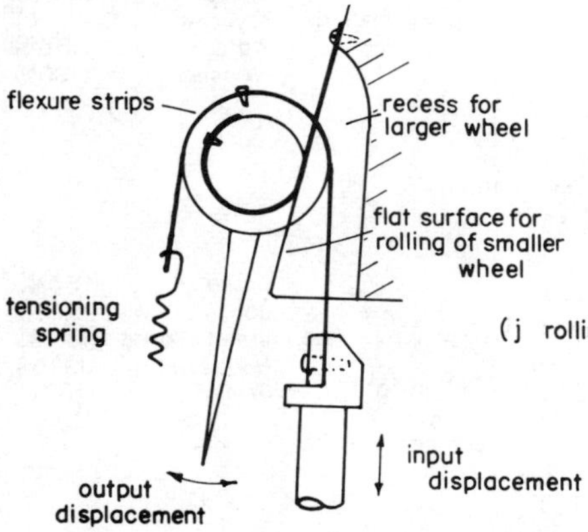

(j rolling suspension

Suggested Reference Sources	
Geary	(1954)
Rolt	(1929)

Figure 5.4 Flat strip springs (continued)

Introduction of a symmetrical but oppositely applied suspension, (Figure 5.4(f)) produces a differential error compensating system that can provide almost pure translation. These are called eight-leaf (or strip) parallel guides. Jones (1968), Trylinski (1971) and Neubert (1975) show two alternative physical arrangements of the compensated method. Jones (1951) and Jones and Young (1956) are basic reading about parallel strip suspensions.

It will be apparent that the suspension produces a reacting force as the table is displaced from the central rest position. The force to deflection ratio is constant over small ranges of travel for both the four- and the eight-strip suspensions. The ratio follows cubic relationships for thickness and spring length and is linear with width of spring material and elasticity modulus. Designs sometimes use two adjacent strips that are merged across into one strip. If the suspension has to bear a downward acting load, the flexures can be stiffened with plates added to the central length of the strip (Geary, 1954).

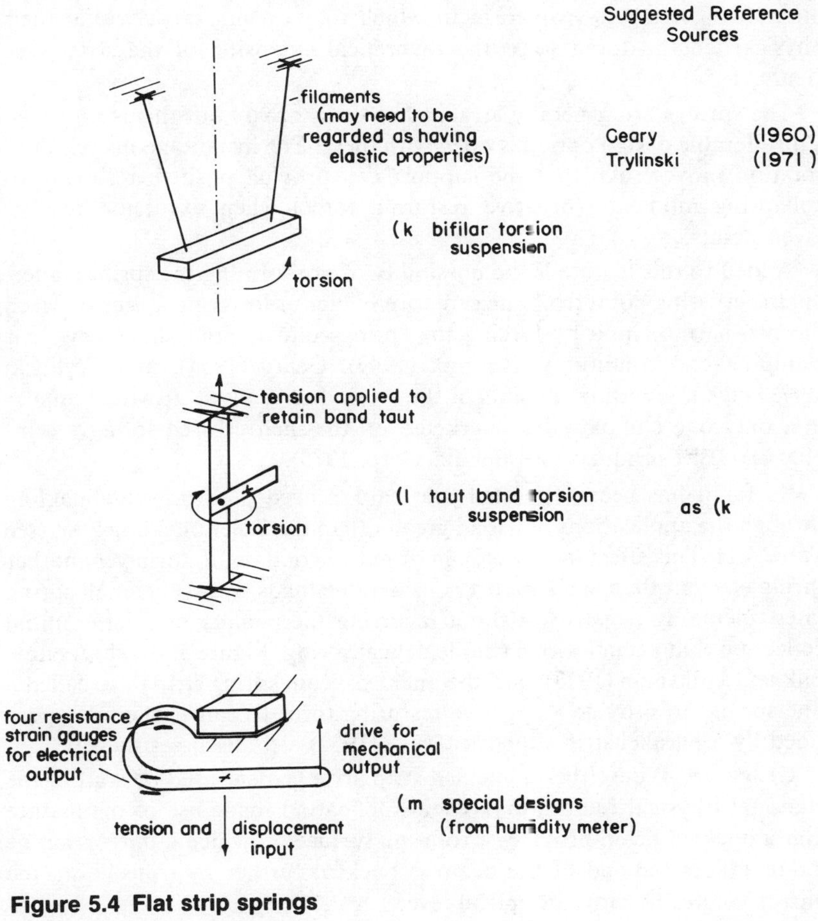

Figure 5.4 Flat strip springs

Parallel strip suspensions will deflect according to the inverse of their stiffness values. Thus, a series connection of a stiff unit driven by a more compliant unit acts to reduce motion in a positive, accurate and stable dividing manner. Figure 5.4(g) shows the layout of strip flexures used to obtain displacement reduction of 1000:1, with negligible friction and force requirements. It was designed to displace an optical grating to a discrimination level of 0.08 μm (Sim *et al.*, 1971). This application is typical of many in which considerable ingenuity is applied to obtain very effective mechanical mechanisms using flexures.

Having replaced the kinematic slide with a flexure equivalent, at least for small amplitudes, it may seem obvious there might also exist a limited action replacement for the rotating bearing. This does exist and is known as the crossed strip pivot (or numerous name variations with 'crossed' used in each). Figure 5.4(h) shows the four-strip hinge. This may take on the alternatives of three strips when two are combined inside the center of the other two, or a two-strip form in which they do not cross within their physical length, doing so at the theoretical extension of the strips (see Geary, 1954).

The springs are generally arranged to cross at 90° but this is not vital. Considerable design option is open with these. The instantaneous center of rotation moves such that the support can provide positive restoring or collapsing moments (negative restoring force) when excursion reaches given points.

Added to this feature is the possibility of pre-forming the springs, altering the crossing point from the center to higher or lower positions between the two surfaces and by forcing the springs into nonflat shapes by their clamping end conditions. Haringx (1949), Geary (1954), and Trylinski (1971) give in-depth treatments of the crossed-strip pivot. It would appear that only one company has marketed off-the-shelf crossed-spring assemblies as OEM products (see Bendix Corp, 1975).

So far it has been suggested that care is needed to prevent buckling through the applications of a load greater than the Euler buckling limit (see Table 4.1). This effect, however, can be put to good use to form yet another spring element than can support considerable loads with very small spring rates (including negative) without incurring the penalty of a large initial deflection that occurs with a simple helical spring. Figure 5.4(i) shows how Eijk and Dijksman (1977) used the buckled-beam spring strip (also called a sine spring) to provide a negative restoring force to cancel the force provided by a parallel-strip support; Geary (1954) also discusses these.

Geary (1954) describes a buckled strip arrangement used by staff of the National Physical Laboratory. Their application made use of the feature that a buckled beam provides a rotating surface on which a mirror can be mounted. As the end of the beam is buckled further by translation, the mirror rotates to form an optical lever.

Strip can also be employed to form a rolling suspension. Figure 5.4(j) shows an example adapted from Rolt (1929) (see vol 1, p 333 for greater detail). The strips here act to hold the roller onto a flat surface while allowing the system one rotational degree of freedom. One spring strip must be tensioned (by a spring or a gravity mass) to keep the system taut. As is illustrated, the linear motion of the input is converted into a rotation that also can be given gain by use of different diameter input and output rollers.

The bifilar suspension has already been mentioned; Figure 5.4(k) shows a form of this. These take numerous forms in practice ranging from the two round section wire filaments shown to systems using multiple filaments. As a first approximation, these can be considered to have completely flexible filaments, but more detailed design would need to consider their elasticity. Trylinski (1971) and Geary (1960) discuss their design in detail; numerous scientific instruments have been built using such suspensions. A similar torsion action can be obtained using a tensioned taut band (Figure 5.4(e)).

Finally, in this short review of the innumerable possibilities it must be stressed that special shapes can often be devised for special purpose use. Figure 5.4(m) shows one case employed in a commercial humidity meter to simultaneously provide tensioning force to a stretched hygroscopic membrane and, as the membrane forces the flexure to bend, to provide load-cell action by driving four resistance-strain gages mounted to its external curved surfaces.

It must by now be clear that flexure elements can play a very important part in instrument design and that kinematic design is not always the better option to choose. If the above is not enough to impress this point, the following section will further prove their widespread utility.

Plate Flexure Elements

Relatively thin plates can be used to provide precision guidance of the center point of the plate; the plate can also simultaneously act to contain pressure of a fluid.

The simplest arrangement (Figure 5.5(a)) uses a flat, thin, flexible sheet of material that is fastened around the periphery either as a fixed or a simply supported edge in which edge rotation is allowed. In the center the member to be guided is fastened. As these provide a rapidly rising restoring force with deflection, they are generally only applicable to guide very small motions. Figure 4.1 in the previous chapter shows their use to guide the armature of a noncontact inductive sensor that needed a working range of only 5 μm. Trylinski (1971) provides expressions that show that those provide a linear force displacement relationship for small movements.

Two basic methods may be employed to obtain a linear force-displacement characteristic over a larger range of movement with a reduced spring rate; that is, those that are more easily deflected.

The first method is to slit the membrane in some appropriate way so that it can deflect further. Figure 5.5(b) shows several possible shapes. As a very approximate guide to estimating the compliance, designs can often be regarded as providing the net effect, summed in the center, of one or more cantilever beams. The spiral case shown can be mentally unwound to form a long thin beam supported only at one end. In some designs the plate is not supported around the periphery but at points within the plate. There appears to be no published analytical design information on their design. Trylinski (1971) and Neubert (1975) devote space to these guides but do not provide analytical expressions for other than the simple continuous plate with point or distributed loading. Gibbs (1957) appears to be one of the few

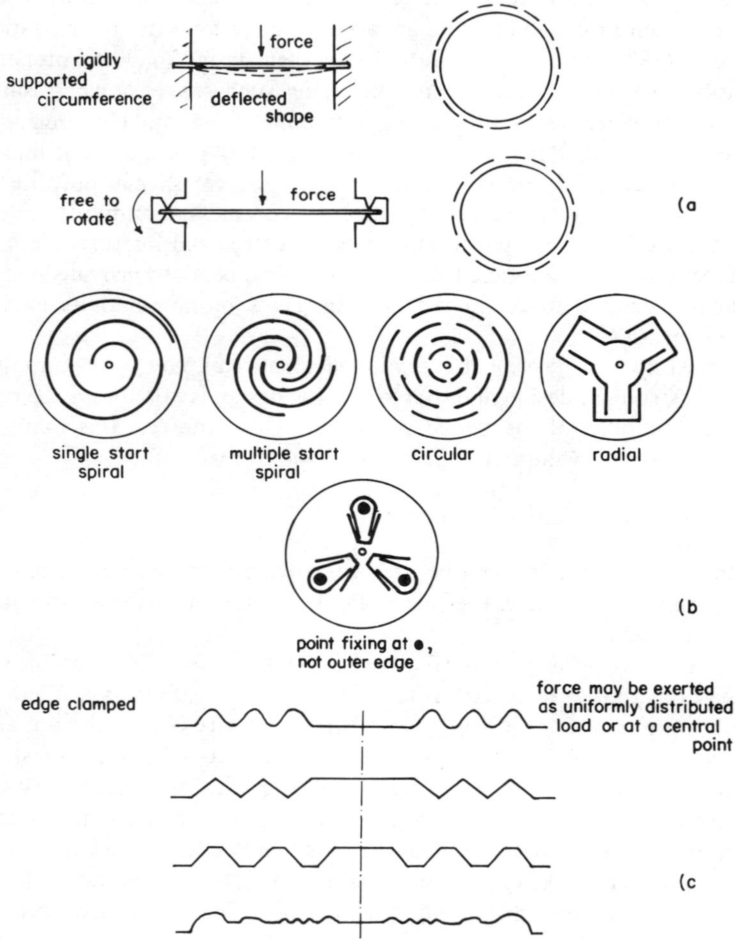

Figure 5.5 Flexure membranes: (a) Solid diaphragms, (b) Slit diaphragms, (c) Corrugated plates and capsules

reports specifically devoted to the subject, but it contains little more than is given in Trylinski (1971).

Thin plates used for this purpose are called slit membrane guides or diaphragm spring guides. They are used extensively in such applications as loudspeaker moving-coil supports, geophone suspensions, vibrator suspensions, and anywhere that relatively small amplitude, frictionless translation is desired. By careful design, they can operate for millions of cycles at high frequency, providing high stiffness across the diaphragm and low stiffness in its plane.

They do not appear to be marketed as a stock item and like most flexures have to be manufactured as needed. Slits are formed by photo-mechanical etching methods, by spark-machining, or by plain stamping where quality does not need to be very high and the production run is large.

Slits may well provide greater compliance and range but they remove the property of the plate to retain fluid pressure. In cases where guidance alone is needed, this may be an advantage to prevent air pressure pumping. The application shown in Figure 4.1, however, deliberately used a closed membrane to exclude moisture and dirt from the inductive sensor.

If pressure-tightness is essential, as it is in pressure gages, some other method of making the diaphragm more elastic is needed. It can be obtained by pressing corrugations into the plate. Several commonly used cross-sectional shapes are shown in Figure 5.5(c).

Sealed diaphragms are also called membranes; Trylinski (1971) calls them pressure controlled elastic elements.

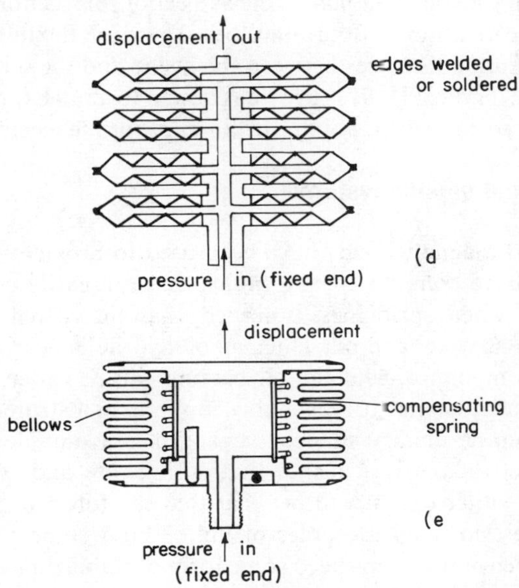

Figure 5.5 Flexure membranes: (d) capsule stack and (e) bellows

Presumably because sealed diaphragms are used so extensively in pressure sensors, a product line that forms a large part of process instrument sales, there has arisen extensive need to pursue their design to considerable depth. Design is reviewed in analytical terms in Trylinski (1971). CDA (1953) provides practical detail of application as do Beggs (1955) and BELL and HOWELL (1974). Computer-aided procedures developed in the Department of Applied Physics of The City University, London have made the design procedures of compliant plates considerably simpler to implement; work previously needing much actual prototype manufacture and testing has been bypassed by the use of rigorous mathematical modelling of the profiles needed.

To obtain greater deflection for a given force, the diaphragms may be joined to form capsules and stacks of capsules, as shown in Figure 5.5(d). Bellows also provide a similar increase in rectilinear flexibility (Figure 5.5(e)). As the characteristics of the materials that must often be used — because of temperature limits and required high strength levels — do not always have the desirable low thermal coefficient of elasticity, pressure systems often need some form of thermal compensation built into the stack assembly. Aneroid barometer bellows and capsules are sealed with internal pressure; the choice of fluid, pressure, and quantity can be made to assist compensation; internal helical springs are sometimes used.

In pressure sensors, diaphragm deflection is transduced by direct linkage to a mechanical device or a sensor might be actually attached to the membrane. In the latter case membrane design must allow for this.

Although not made from plates, the subject of this section, it is appropriate to include mention of Bourdon tubes. These are flexible tubes formed such that pressure differences between the inside and the outside cause the tube to deflect. Trylinski (1971) provides design formulae for these. Deflection is harnessed to drive a pointer or an appropriate electric sensor.

Nonmechanical elastic systems

Electric and magnetic fields have been used to provide variable spring rates and so have columns of air. Fluids that are easily compressed are more suitable (where springiness is needed) than the virtually incompressible fluids such as water and oil. Thus, an oil hydraulic bearing is very stiff, having little compliance, while an air bearing will be softer.

Magnetic suspensions are reasonably common in instruments. They can be formed from permanent magnets arranged in a suitable manner using additional electromagnetic coils so that spring rate and stability can be controlled. In either case it will be clear that the force of attraction rises sharply as the two magnetic poles of unlike kind come closer together. Magnetic suspension systems need some form of stabilizing arrangement to make them statically stable. Earnshaw, in 1839, showed theoretically that the inverse square law for force and distance fields (magnetic and electric fields are such) cannot inherently support a fixed magnitude pole in a stable manner (Geary, 1964). Automatic control is employed to alter the force with distance as a closed-loop system. Further stabilizing mechanisms may be needed to ensure that one pole does not slide away from the desired position. This can be visualized when attempting to sit (levitate) a pole over the top of another pole of like polarity.

An advantage of magnetic springs is that they can be made to act through sealed walls of containers. An example application might be as the connection and weighing medium for a sensitive chemical weighing balance. The process of measurement can be achieved by altering the field strength of an electromagnet so that the balance returns to a given datum in a manner depicted in Figure 5.6. The electric current needed to achieve the balance state is related to the weight of a mass supported.

Electric fields also exert forces so they can be used as elastic elements in the design of an instrument. This method has been used in at least one gravimeter to restore the balance position.

Further detail on magnetic and electric suspensions is to be found in Geary (1964). Weighing balance applications are detailed in Gast (1968), Gast (1974), and Wolsky and Zdanvk (1969).

As the name implies, an air spring is a compliant movement obtained by the use of air, contained in some way, to support a given load. The automotive pneumatic tire is the most obvious example of air spring, being used in that application to assist isolation of vibration.

Air springs are capable of substantial load bearing without large deflections, and they can easily provide low natural frequency, second-order suspension systems. They have been used extensively in vibration isolation of machinery. The spring usually takes the form of a round flat flexible bag that is pumped up to a suitable pressure.

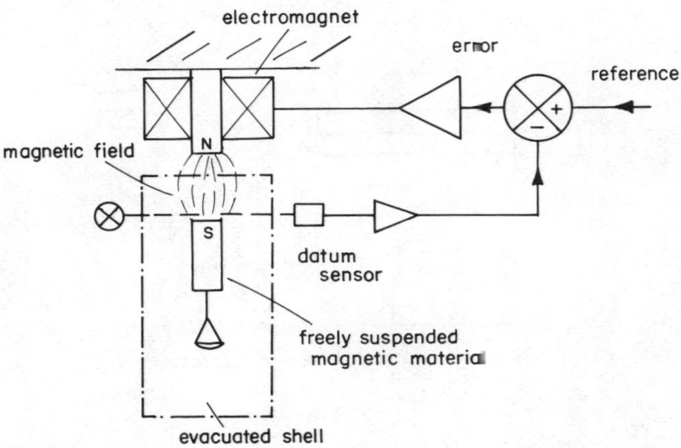

Figure 5.6 Schematic of force-balance, magnetic suspension weighing technique

Air springs have no resonant effects within the spring itself (as do mechanical coil springs). Their rate can be made to stiffen with increasing load, being nonlinear devices. This means that second-order systems formed with them remain at substantially the same natural frequency with changing load. They are, however, affected by ambient temperature, and they require a relatively large amount of maintenance work. They are generally regarded as being suitable for only heavy load situations. The reader should consult Jones (1967) and Dorien-Brown (1971) for further references and an introduction to this form of elastic system.

Bimetal and biomorph systems

If two materials (Figure 5.7(a)) having different thermal expansion rates are placed side by side with one end of the parallel combination fixed securely, temperature changes to the element so formed will bend traversely due to differential expansion between the two elements. Such elements are called bimetallic devices if made from metals. A similar device can be formed (Figure 5.7(b)) using two pieces of piezoceramic that are fastened together and have deposited electrodes placed along their outer edges and the central interface. These are termed bimorph-benders, or twisters.

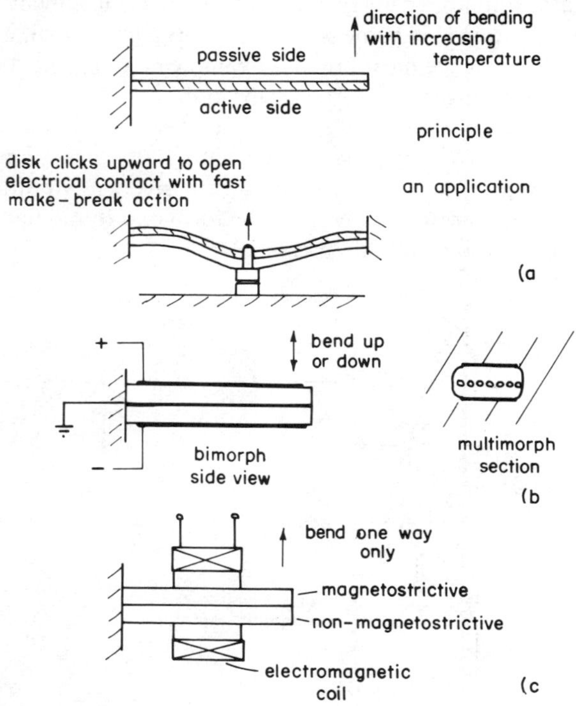

Figure 5.7 Differential expansion bending and twisting devices: (a) Bimetals, (b) Ceramic bimorph and multimorph, (c) Magnetostrictive bimorph

Bimetals bend with temperature, converting temperature change into displacement. Bimorphs convert electrical energy into displacement. Bimorphs involve a reversible process — that is, deflection will generate electricity — but bimetal elements are irreversible.

Bimetals are made by soldering or welding thin plates of suitable material together, the final shape being stamped as strips, disks, or whatever shape is needed. The side having the higher thermal coefficient of expansion is called the active side; the reverse is known as the passive side. As no material has an absolutely zero expansion characteristic and metals generally possess only positive coefficients, the net effect is always that of a combination of two positive coefficient expansions. The forces of this expansion are considerable, and bimetals elements can often directly drive an electrical contact or dial indicator.

To obtain the greatest deflection effect, the metals used need to be, for the passive side, one with the lowest practical coefficient (which is usually Invar) and on the active side one with a high coefficient (stainless steels, brasses, and special alloys).

Traverse deflection can be converted into rotation through the use of spiral spring forms. These also can be used to obtain greater deflection. However, as the deflection per unit thermal change parameter is increased it will usually be found that the stiffness of the bimetal element reduces, making it less capable of driving a load without distortion. Preloading the bimetal element against a stop (prop) provides a means to adjust the datum temperature position at which the element begins to respond.

The design of bimetal elements is covered in detail in Timoshenko and MacCullough (1959) and in Trylinski (1971).

Bimorph devices are less common than bimetal elements and information about them is a little harder to locate. Neubert (1975) gives a short account of how piezoelectric sandwiches can be made to provide bending, twisting, and beam deflection action, as well as bending a plate across the diagonal or lifting the free corner of a three-corner mounted plate.

Design information is usually available from the suppliers of the materials, Brush-Clevite being one source of application notes.

A variation of the bimorph is the multimorph. These can provide greater elasticity and are said to be easier to manufacture. Instead of sandwiching two strips together, a single strip is made with a row of fine holes formed through the center line. These are filled with conductive material to form the inner electrode. The outer faces are coated with electrode material in much the same manner as for the true two-slice sandwich bimorph.

Other forms of differential bending devices could be formed on this principle but they do not appear to have been used. For example, a nickel strip sandwiched (Figure 5.7(c)) to a stainless steel side would bend in the same direction with an increasing magnetic field of either polarity; the nickel, but not the stainless steel, is magnetostrictive. This is the magnetic equivalent of the ceramic bimorph. It would also give reversible characteristics.

Bimetal and bimorph devices can be used to provide compensation for the physical attribute that suits their input energy form. Bimetal elements are often used, for example, to correct diaphragm-type pressure gages. They are also used to trip circuits when the ambient temperature produced by the direct true ambient or by the use of a heating coil or direct heating causes the strip to deflect. Chapter 7, on error sources and their reduction, provides more information about the use of thermal differential methods.

The dynamic response of such differential devices may be important. Bimetal devices are relatively slow to act since they must transfer thermal energy to obtain deflection. They will be substantially first-order systems in response because the thermal storage and dissipation rate combination is much slower than the rate at which the strain energy transfers. Clearly, the smaller the element, the faster the response, and physical size limits would generally provide these elements with time constants of seconds order. Miniature thick and thin film devices are able to respond much faster but at the expense of less power drive capability.

Bimorphs are inherently high frequency devices. As the rate of energy change for the electrical and the strain energies transfer is of the same order, these will provide second-order response with high natural frequencies in the kilohertz and higher region; low damping is a feature. They are not, of course, a replacement for the bimetal since the two devices operate with different input energy forms.

References

Andreeva, L. E. 1966. *Elastic elements of instruments*, Translation from Russian, Israel Program for Scientific Translations Ltd, Israel.

AMETEK. 1973. *Design data for Negator constant force springs*, Hunter Spring Div., Hatfield, PA, USA.

Baird, K. M. 1968. 'Compensation for linear thermal expansion', *Metrologia*, **4,** 145–6.

Beggs, J. S. 1955. *Mechanism*, McGraw-Hill, New York.

BELL and HOWELL. 1974. *The Bell and Howell pressure transducer handbook*. Bell and Howell, CEC/Instruments Div, Pasadena, USA.

Bendix Corp 1975. *Free flex flexure pivot engineering data*, Bendix Corp, Pub No 00U-6-752B, Electric and Fluid Power Div, New York.

Braddick, H. J. J. 1954. *The physics of experimental method*, Chapman and Hall, London.

Carson, R. W. 1937. 'Measuring elastic drift', *Proc ASTM*, **37,** 661–675.

CDA. 1953. *Copper in instrumentation*, Pub No 48, Copper Development Assoc, Radlett, UK.

Chironis, N. P. 1961. *Spring design and application*, McGraw-Hill, New York.

Dorien-Brown, B. 1971. *Air springs for vibration isolation*, Harold Armstrong Conf on 'Production science in industry', Institution of Engineers, Australia, Melbourne, August. 213–225.

Dratler, J., and Block, B. 1972. 'A wide-band horizontal accelerometer with preliminary Earth Normal mode and seismic investigations', *Geophys J R Astr Soc*, **27**, 337–367.

Eastman, F.S. 1937. 'The design of flexure pivots', *J Aeronautical Sciences*, **5**(1), 16–21.

Eijk, J. van, and Dijksman, J. F. 1977. *Plate spring mechanism with constant negative stiffness*, Internal Report, Fine-mechanics Group, Technische Hoschule, Delft, Netherlands.

Eisley, J. G. 1963. 'Nonlinear deformation of elastic beams, rings and strings', (Bibliography, 92 refs), *Applied Mechanics Reviews*, **16**(9), 677–679.

Faires, V. M. 1955. *Design of machine elements*, Macmillan, New York.

Frazier, R. H., Gilinson, P. J., and Oberbeck, G. A. 1974. *Magnetic and electric suspensions, types of suspensions developed in the Charles Stork Draper Laboratory*, MIT Press.

Fremery, J. K., and Doden, K. 1978. 'Active permanent magnet suspensions for scientific instruments', *J Phys E: Sci Instrum*, **11**(2), 106–113.

Gast, T. 1968. *Conversion of measuring values for mass, density and flow into proportional frequencies with the aid of free magnetic suspension*, In IEE Conf Pub 43, *Industrial measurement techniques for on-line computers*, London, June.

Gast, T. 1974. 'Vacuum micro balances, their construction and characteristics', *J. Phys E: Sci Instrum*, **7**, 865–875.

Geary, P. J. 1954. *Flexure devices — pivots, movements, suspensions*, Sira Institute, Chislehurst, UK.

Geary, P. J. 1960. *Torsion devices*, Sira Institute, Chislehurst, UK.

Geary, P. J. 1964. *Magnetic and electric suspensions*, Sira Institute, Chislehurst, UK.

Gibbs, D. F. 1957. 'Spring diaphragms', *J Sci Instrum*, **34**, 34–35.

Gohn, G. R. and Fox. A. 1965. 'Stress relaxation in metallic springs and conductors', *Electrotechnology*, **16**, 85–92.

Haringx, J. A. 1949. 'The cross-spring pivot as a constructional element, *Appl Sci Res*, **A1**, 313–332.

Hugill, A. L. 1978. *Gravimeter design*, Int Rep DSS/AH/171, Systems Science Dept, The City University, London. September.

Jones, R. V. 1951. 'Parallel and rectilinear spring movements', *J Sci Instrum*, **28**, 38–41.

Jones, R. V., and Young, I. R. 1956. 'Some parasitic deflexions in parallel-spring mounts', *J Sci Instrum*, **33**, 11-15.

Jones, R. V. 1967. 'The measurement and control of small displacements', *Phys Bull*, **18**, 325–326.

Jones, R. V. 1968. 'More and more about less and less', *Proc R Instn Gt Br*, **43**(202), 323–45.

Minato, K., Nakafuku, C., and Takemura, T. 1969. 'Rolling friction of metals', *Jap J App Phys*, **8**(10), 1171–1179.

Neubert, H. K. P. 1975. *Instrument transducers — an introduction to their performance and design*, Clarendon Press, Oxford.

Roark, R. J., and Young, W. C. 1975. *Formulas for stress and strain*, McGraw-Hill, New York (Editions since 1938).

Rolt, R. H. 1929. *Gauges and fine measurements*, 2 Vols, Macmillan, London.

Schmidt, R., and DaDeppo, D. A. 1973. 'Variational formulation on non-linear equations for straight elastic beams', *J Indust Maths Soc*, **23**(2), 117–136.

Sim, P. J., Ross, H. A., and Taylor, R. J. 1971. *Photographic gratings for linear measurement*, Harold Armstrong Conf on 'Production science in industry', Institution of engineers, Australia, Melbourne, August, 57–68.

Sydenham, P. H. 1969. *Manual and automatic large-scale dimensional metrology*, PhD Thesis, University of Warwick, UK.

Sydenham, P. H. 1982, 1983. *Handbook of measurement science*, Wiley, Chichester.

Sydenham, P. H. 1984. "Elastic design of fine mechanism in instruments" *J. Phys. E.: Sci. Instrum.*, **17**, 922–930.

Thorpe, A. G., II 1953. 'Flexure pivots-design', *Product Engineering*, **24**(2), 192–200.

Timoshenko, S., and MacCullough, G. H. 1959. *Elements of strength of materials*, Van Nostrand, New Jersey (Numerous editions since 1935).

Timoshenko, S. P., and Gere, J. M. 1961. *Theory of elastic stability*, McGraw-Hill, New York.

Tovey, F. M. 1974. *Material properties in the design of transducer flexibles*, Ai Research Manuf Co, Phoenix, USA (plus illustrations to accompany above as separate).

Trylinski, W. 1971. *Fine mechanisms and precision instruments — principles of design*, Pergamon Press, Oxford, translated by A. Voellnagel from Polish original.

Votta, F. A., Jr. 1952. 'The theory and design of long-deflection constant force spring elements', *Trans ASME*, **74**, 439–450.

Wahl, A. M. 1963. *Mechanical springs,* McGraw-Hill, New York.

Weinstein, W. D. 1965. 'Flexure pivot bearings', *Machine Design*, Part 1 June, 151–157, Part 2 July, 136–145.

Whitehead, T. N. 1934. *Instruments and accurate mechanisms — Underlying principles*, Macmillan, London (Also Dover Pubs, New York, 1954).

Wolsky, S. P., and Zdanvk, E. J. 1980. *Ultra-microweight determination in controlled environments*, Wiley, New York.

Chapter 6
Mathematical elements

The relevance of mechanical mathematical mechanisms

With the introduction of digital computing methods based upon electronic solid-state active devices, it might be thought that mechanical computing has no longer any part to play in instrumentation; this is not so. Often mechanical computation may be a more economical and applicable solution. Mathematical operations are an important part of the capability available in the mechanical design of instrumentation.

Mechanical mathematical mechanisms are those in which the mechanical input signal bears a known and adequately held common mathematical relationship with the output signal. They are a special class of mechanism in which normal precision design complements additional requirements for mathematical operation.

They largely find competitive application with electronic alternatives in situations where the number of function relationships required is relatively small and where electrical power is not readily available for electronic data processing circuitry. In this case the mechanical input to the instrument provides the energy to actuate the mechanism. The break-even point at which the electronic alternative is superior to the mechanical solution will generally occur when anything more than the simplest of mathematical operation is needed. Mechanical methods are particularly suited for the design of compensating mechanisms needed to correct an inadequate primary function formed by a mechanical system.

In some applications the mechanical method can provide greater reliability. Mechanical mechanisms generally fail to give proper operation through gradual deterioration of the accuracy, which steadily falls with wear. Electronic equivalents, unless built with redundancy arrangements, will generally fail absolutely and often catastrophically.

Mechanisms retain absolute position, thus storing information. This feature can make them suitable for applications where past history has to be stored over very long periods. An obvious example would be the use, in a remote location, of the mechanically driven chart recorder such as that shown in Figure 6.1, in which, say, pressure head is converted into a velocity related signal from a differential pressure form of flow sensor.

Unlike electronic systems where signals must be converted back into positional movements before the data resulting from the mathematical operation can be seen, a mechanical system immediately shows the state of the system. For this reason they are often more easily accepted by users.

The advantages of mechanical computing, however, can often be largely offset by those disavantages that brought about its general demise beginning in the late 1950's. Figure 6.2 is of an early tide predictor (by Kelvin) that calculated a Fourier series function.

These systems are generally relatively large and heavy. They require good maintenance to ensure that they operate efficiently and without wear. They are costly to manufacture and they do not assemble easily from basic building block functional units. For these reasons mechanical methods of computing tend to be applicable only to instruments needing a very small amount of mathematical manipulation.

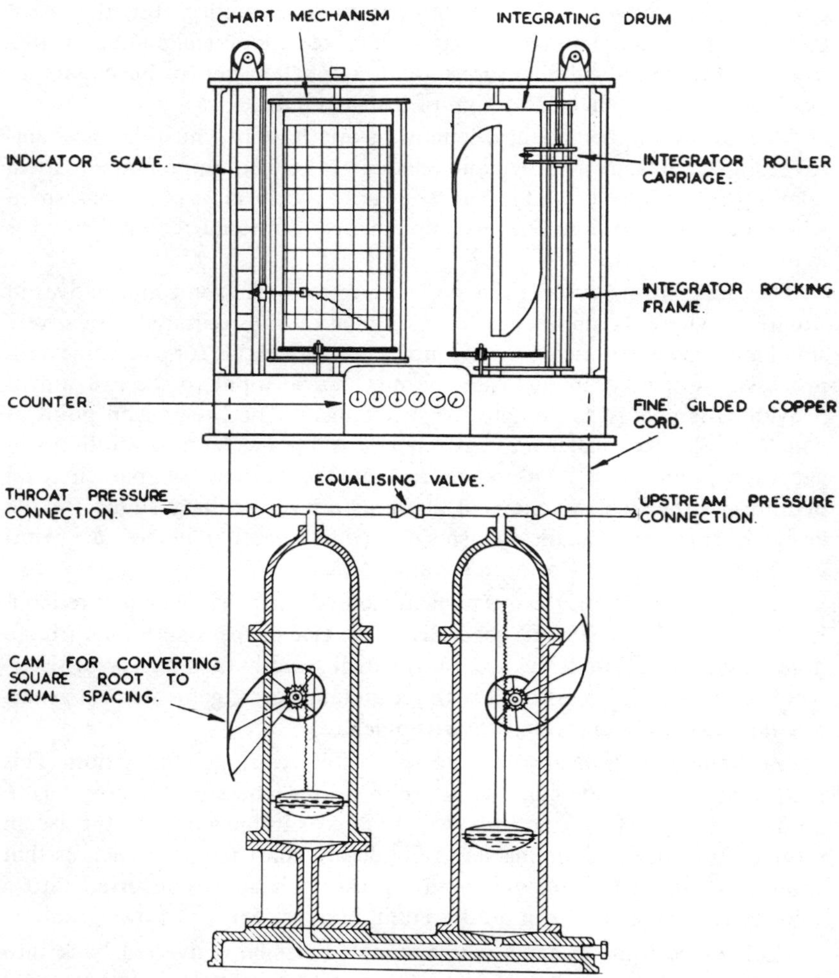

Figure 6.1 Early Kent flow recorder using cams to convert input data into a linear form (© Brown Boveris Kent. Used with permission.)

Applications

To illustrate when the mechanical solution may be favorable, it is instructive to consider applications that would, today, still employ a mechanical rather than an electronic method. If the instrument has to be extensively developed in terms of its information processing role at the original design stage, it is most likely that an electronic system will be the best choice. From that point on it is most unlikely that a mechanical counterpart can be justified.

For example, the first application considered is for a means to apply automatic focusing in enlargers and cameras. Simple equipment will find the use of appropriate mechanical linkages a very convenient method. If the system needs, however, to operate at the higher level of sophistication for

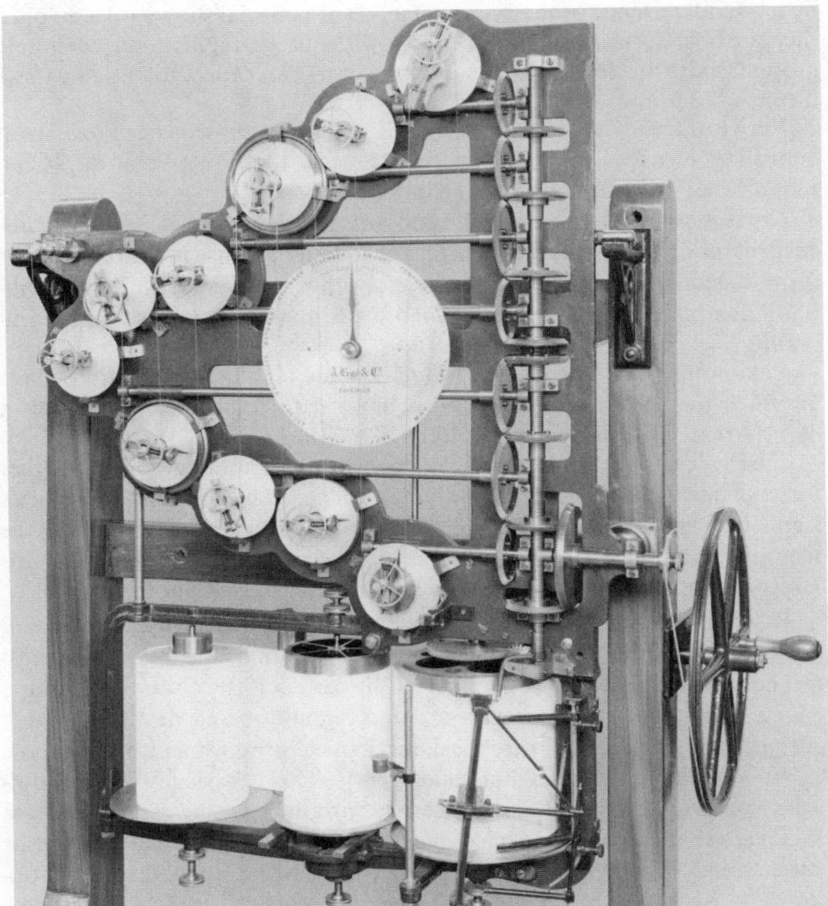

Figure 6.2 Kelvin tide predictor, c 1875 (© Science Museum, London, UK. Used with permission.)

such apparatus, the extra data processing power called for will warrant an electronic processing system. In this case it will usually be more economical to use as little mechanical componentry as is possible. The designer must decide from both initial and projected market design needs (increasing sophistication as the market is widened) which path to follow. If there is a strong likelihood that the system will be later expanded in its data handling areas, then to opt for the electronic solution from the onset may be more sensible. It may, however, carry a price penalty at the early stage.

Other examples are to be found in several forms of surveying instruments. One kind uses mechanisms to perform the computations needed to reduce automatically measuring staff sight readings into range values.

The design of planimeters is another instance. These devices are used to measure the area bounded by an irregular curve, such as that represented by a boundary fence or the profile of a pile of material shown on a map. In the portable form these will be very difficult to replace, on economic grounds, with an electronic version; may users can expect to be using these far into the future.

Very low frequency function generators, such as those needed to control annealing furnaces that operate over a cycle time of several days, can economically use mechanical function generators.

Correction mechanisms in ruling engines and other specialized mechanical generating apparatus may be simpler to build and use in the mechanical form, especially when they must operate with a requirement of 100% reliability over some 15 days as is needed to form a good quality diffraction grating.

Yet another example is to be found in some high quality record players in which the stylus is made to move inward to track the record with a defined tracking function as the limited length are moves inward.

Finally, consider building a speed control for a small gasoline or other form of inexpensive engine. The use of a mechanically derived velocity signal plus spark advance control with speed is a reasonable solution in the face of the common lack of suitable electric power source and the large levels of vibration, heat, and electrical interference that are present.

It will be seen that each of the functions given above requires analog forms of computation. Although digital calculation can be performed with mechanical methods (Babbage proved that in his 19th century calculating engines; many statistical processors, and digitally based desk calculators (Beggs, 1955) were made after that), mechanical arrangements cannot now compete with electronic digital counterparts. This is basically because digital systems are fundamentally concerned with the extreme use of two-state logic cells that are interconnected in giant proportions. Mechanical systems suffer greatly from the overuse of connections, which introduce wear, friction and backlash. Mechanism is far more suited to systems having few bearings, and to those designs where considerable information is held by single continuous components. For example, a lever arm ratio-converting mechanism can provide a thousand positional states over its range, each being a stored value; furthermore, display of these states as relative position is inherently embodied.

Another area where the mechanical computation method still is relevant can occur where it interfaces another energy regime. The logarithmic electrical potentiometer forms the required mechanical rotary linear input to electrical logarithm output by suitable mechanical use of materials in the electrical resistive track. Shutters in some cameras make use of tapered profile apertures to obtain the desired exposure for time-of-opening relationship.

Rigorous design of mechanical computing systems

The rigorous design of mechanical computing elements is covered in several works; see principally Beggs (1955), Mabie and Ocvirk, (1958), Murray (1948), and Svoboda (1948). Having decided the form of the system, the design of actual components within it can make extensive use of Trylinski (1971) and other references given elsewhere in this work. Key mathematical foundation design work for some forms of mechanism used was laid in the last century (Thompson and Tait, 1883). Eames and Eames (1973), Pugh (1975), and Sydenham (1979) each can be used to search for design arrangements already used.

Systematic design begins with determining the mathematics of the computing operations needed, their range, and relevant factors such as initial conditions, speed of action, and adjustments that will be needed within the system.

Important considerations are the energy loss of the mechanism, the input power available to drive it and needed by the mechanism, and the output power required to operate the next device. In electronic systems these considerations generally are easily altered if a design error is made. In mechanical systems it is far more difficult to tune the system, and it is therefore most important that initial design studies consider these factors in considerable detail before manufacture begins. Mismatched stages, in which power and positional requirements do not interface correctly, may result in the need for extensive redesign.

The computations needed can be broken down into serial and/or parallel paths of possible mechanism. These paths can, in turn, be broken down into the basic computational elements that are possible with known mechanical linkages and other mechanisms. Several of these are detailed later to assist initial design. Murray (1948) devotes a chapter to the solution of problems, giving guiding principles.

For each mathematical function there are usually several alternatives available. The designer must consider each in terms of the parameters and the task. Beggs (1955), Mabie and Ocvirk (1958), Murray (1948), and Svoboda (1948) each provide mathematical explanations of the commonly used mechanisms.

From then on, design follows that for any precision mechanical element. Procedures of synthesis, as opposed to analysis of known arrangements, may be more profitable in this case. Many design situations provide the mathematical relationship required from largely empirical studies in which the fundamental mathematics involved may not be formulated.

The block diagram approach, so well used in electronic design, is an admirable method for beginning a design. At that stage it is often possible to arrive at more than one arrangement for solving a given equation. One of these may turn out to be more suited to the specific application. Mabie and Ocvirk (1958) expand this point with an example. (It is historically interesting to note that their work, written in 1958, explains how analog mechanical computers are superior, by far, to the electronic digital systems of that time. Quite the reverse is true today!) Beggs (1955) provides several worked examples of actual designs. They begin with the physical situation from which are developed the mathematical relationships needed. These are then transformed into practical linkages and mechanisms. It will often be necessary in mechanical systems to make use of additional mechanisms that enable the size and range of operation of the respective linkages and parts to be transformed to keep the movements within given space limits.

Facilities will also often be needed to provide conversion from one plane to another and from rotation to linear action. The counterparts of such difficulties are not met to the same extent in electronic systems. In some cases the actual mathematical relationship needed may never be formulated, actual calibration and adjustment being used to arrive at the desired corrective mechanism characteristic. Cams and adjustment corrector plates are examples where this occurs. The inertia of the mechanism must be carefully studied to keep wear and operative speed within desired limits.

Basic function elements

Addition and Subtraction

Several alternative mechanisms are available for adding or subtracting two mechanical input variables. The inputs may be in linear or rotational form, each needing the appropriate 'adder' mechanism.

A first method makes use of the principle that two pulleys, as shown in Figure 6.3, moved in accordance with two input displacements A and B, will pull out a tensioned chain, tape, or cord a distance equal to twice the sum, or the difference, of A and B. This concept can be applied in a variety of ways to add a virtually unlimited number of addends. The input directions can be varied in direction and location. Murray (1948) discusses an arrangement wherein an endless band can be used instead of one as shown that needs constant tensioning at one end. This method was used by Thompson and Tait (1883) to sum Fourier components in their tide predictor shown in Figure 6.2.

Rigid bars that are constrained to slide in a parallel manner can be used to add two variables (see Figure 6.4). The output displacement at the center of the cross linkage will, if the three move parallel, be equal to half the sum of the two A, B inputs. Clearly the design of the slides and the length of the three rods must be chosen to suit the displacement range of the inputs and output. Several alternative layouts are illustrated in Mabie and Ocvirk (1958) and in Murray (1948). Some designs are not strictly accurate adders, making use of the sin $\theta = \theta$ approximation for small angular excursions.

Addition of linear inputs can also be obtained using two racks that sandwich a central rolling disc. The system may be friction driven or make use of more positive linear gear tracks. This option can provide a very slim adding mechanism. The abovementioned works contain further information.

Addition or subtraction of rotations can be obtained using driven wheels in the manner of Figure 6.3, where A and B are rotary inputs rather than linear translations. Movement of the tape or chain will be equal to the sum

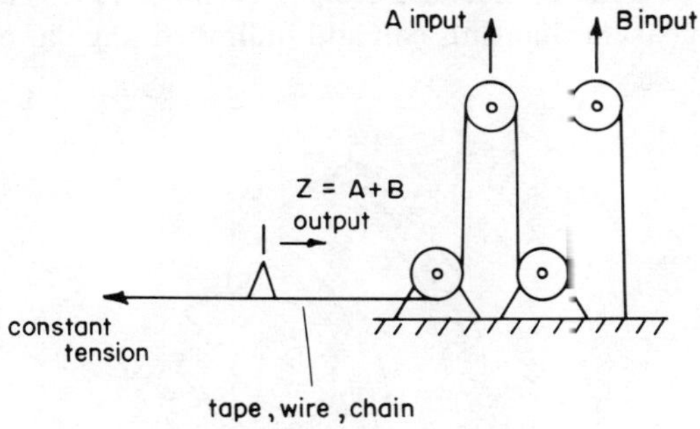

Figure 6.3 Addition using pulleys and a tape

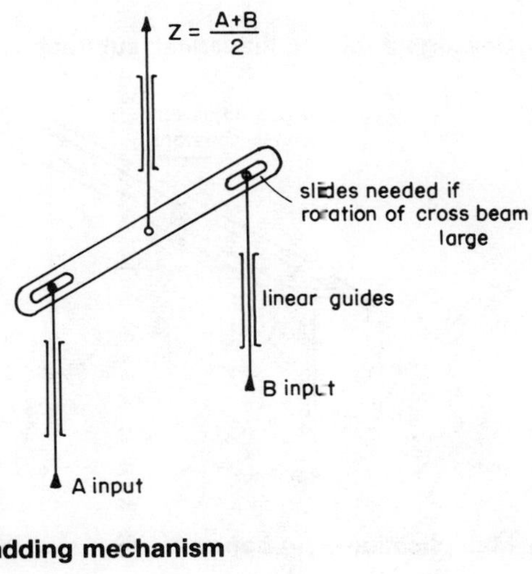

Figure 6.4 Rigid bar adding mechanism

of the two rotations of the input wheels. In this case it is important that the drives possess no slippage. The chain must also be free to move as would be the case for an endless chain system. Furthermore, the inputs A and B must be applied in serial and then be released.

The more usual and satisfactory method is to employ one of the several forms of differential gearing systems that are available. These are rotary equivalents of the linear rack adder. The most commonly met form is that using bevel gears; the method used in the final drive of motor cars to allow the drive wheels to rotate at different speeds when cornering. Figure 6.5 shows the layout of a common instrument differential unit using bevel gears. They are available in very small sizes expressly for mechanical computation.

Alternative forms use spur gears in a somewhat similar manner to the above or as an epicyclic layout. More detail is given in Mabie and Ocvirk (1958) and in Murray (1948). The design of differential gear trains is usually discussed in texts on general machine design. Trylinski (1971) also is relevant. Differential gear mechanisms can add unlimited angular excursions.

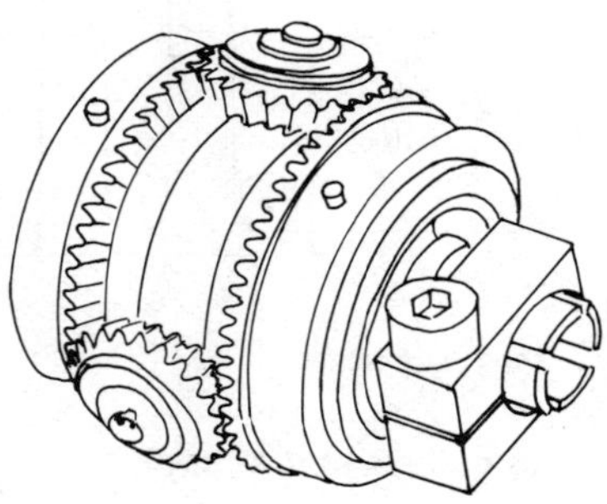

Figure 6.5 Bevel-gear, differential adder/subtractor

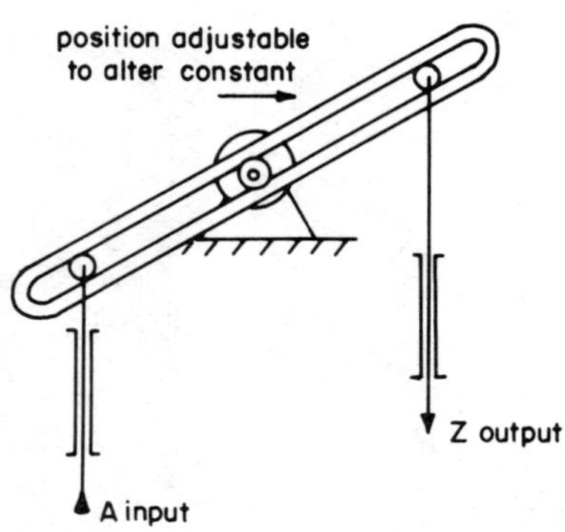

Figure 6.6 Multiplication-by-a-constant, lever multiplier

Multiplication and Division

The simplest form of multiplication between two inputs occurs when one is a fixed quantity. This is the case of multiplication by a constant.

A simple lever, or gear train in the rotational case, provides constant multiplication by a factor set by the ratio of the two lengths of the arms (or gear ratio) of the lever. The lever may be arranged as a centrally pivoted device (which provides reciprocals, being an inverter) or as an end-pivoted bar. In the latter case the input signal is not inverted in direction of movement.

A simple lever, however, only provides accurate multiplication if the angular excursion of the arm is small enough to allow the sin $\theta = \theta$ approximation can be made accurate if the pivot points are constrained to move only in parallel straight paths as shown in Figure 6.6. A centrally pivoted constant multiplier is shown.

Movement of the position of the pivoting point for the linear system, and gear ratio changing in the rotational counterpart, allow for the constant factor to be varied. Where the rotations involved require only part revolution motion, it is permissable to use gear sectors.

Multiplication of two varying inputs is more complex to perform. The most commonly described methods make use of sliding linkages and pivoting points of links to set up multiplication based on the theory of similiar triangles. Figure 6.7 shows the layout. Clearly, the manufacture of this mechanism can be expensive if accuracy is required. If such an operation is needed, it may well be that it is more expedient to convert the linear displacements into electrical equivalent signals and then use electronic analog multiplier modules.

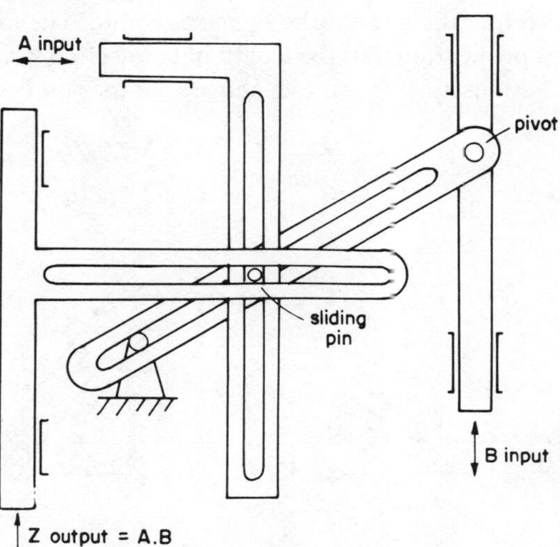

Figure 6.7 Similar triangles multiplier

There is, however, a simpler multiplier that may suffice; it is not an exact mechanism. This is called the links multiplier and is based on the sin $\theta = \theta$ approximation. It is discussed in Murray (1948).

Multiplication is also possible using the relationship:

$$AB = \frac{(A + B)^2}{2} - \frac{(A - B)^2}{2}$$

Here the complexity of accurate multiplication mechanism is replaced by less severe addition mechanism plus two squaring stages. It is also possible to conduct multiplication using logarithm methods as explained in Murray (1948). In general, multiplication will require expensive mechanical systems.

Squares, Square Roots, Square Roots of Products

Several methods are available for producing either of the variations of squaring, square rooting, and providing the square root of a product. That shown in Figure 6.8 is of the form shown in Mabie and Ocvirk (1958). If the two inputs A and B are fed in as translations of the two A and B bars, the vertical bar Z will move as the square root of the product of A and B. Further squaring combinations are obtained by feeding into Z as one input.

If the input shaft B is held fixed, the output shaft Z will be the square root of input A, multiplied by a constant factor set by the position to which shaft B is displaced and held.

A second method, one that is capable of high precision in producing the square of an input shaft motion, is that marketed by Librascope; for illustrations see Beggs (1955) and Mabie and Ocvirk (1958). An input shaft drives a cone supported on the same shaft. The cone drives a parallel mounted cylinder through a cable arranged such that it remains taut despite the change of center distance as the two axes rotate. The radius of the cone at any point is proportional to the length of connecting string to one side of that point. Consideration of this fact shows that the rotation of the cylinder

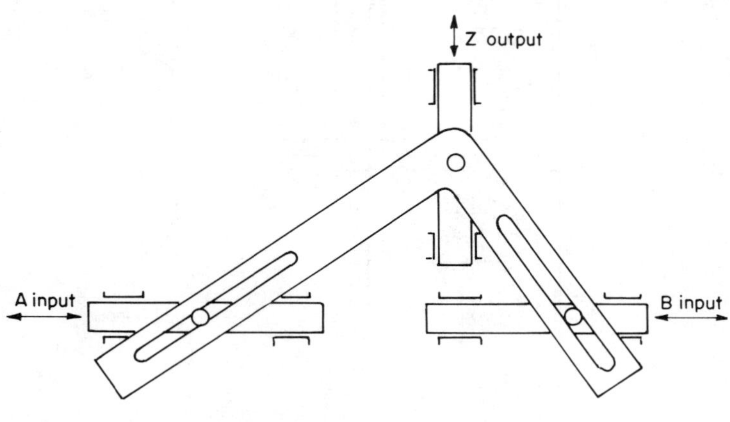

Figure 6.8 Sliding-bars squaring mechanism

is thus the square of the rotation of the cone. To make practical device — the cone cannot run to a zero point in the layout shown — it is necessary to add in a constant to the input quantity and then extract the output using a subtraction differential. The wire connecting drive can be replaced by an idling wheel if the cone is set to be such that its surface lies parallel to that of the cylinder.

Mabie and Ocvirk (1958) and Murray (1948) each describe yet one more method for obtaining squares: that using contour cams. These are cams that run in constant contact without slipping. Gear teeth are often used to ensure this. Given fixed centers of rotation for the two meshing cams, it can be shown that rotational output of one is the square of the other input. Reverse operation provides the square root. When light drive torques are involved, it is possible to make use of smooth-faced friction cams.

Contour cams can also be used to generate other relationships between input and output. They can also be used to generate logarithms, which, in turn, can form the basis of multipliers when used with adders. The flow recorder shown in Figure 6.1 makes use of shaped cams to linearize a square root prime input function.

Function Generation

Generation of the simple trigonometrical sine and cosine functions can be achieved using mechanism in which the projection of the circumference of a turning circle is obtained at a linearly constrained point. The Scotch yoke mechanism is one device used for this.

An alternative method is based upon a planetary gear system in which the degenerate hypercycloid function is utilized to form simple harmonic motion.

Tangent relationships between input and output mechanical motions are as easily obtained. Beggs (1955) lists other methods of trigonometrical function generation.

More complex functions can be generated using appropriately shaped cams provided they do not require too rapid a function value change. Linear cams may be more suitable then rotary forms for this application.

Integration

Many instruments require the primary input data to be integrated. Examples are energy meters for gas and electricity, and water flow recorders. In addition to these, the need for mathematical computation concerning differential equations will need integration.

The most common mechanical integrating device is known as the disc, ball, and cylinder integrator. This was invented by Lord Kelvin's brother James Thomson around 1876. Figure 6.9 shows a contemporary etching of his mechanism. Modern forms remain largely identical with this.

The disc D is rotated according to the magnitude of the input variable. This rotates the cylinder via the transferred motion of the ball that is assumed to be rolling in a pure rolling motion. Modern systems generally make use of two balls in series instead of one, the two being held in a carrier in which the position of the balls can be set across the cylinder. This is not shown in the Thomson's etching, but is needed in practice. The rotation of the output cylinder is the integral of the input signal with the respect to the other input, which is the position of the balls across the cylinder.

If the position of the balls is held constant, the unit acts to produce multiplication of one input by a constant. According to Mabie and Ocvirk (1958), commercial units can yield integration to within 0.01%. Thompson and Tait (1883), Murray (1948), and Trylinski (1971) are prime references about this mechanism.

The disc, ball, and cylinder mechanism can also be used as a variable ratio multiplier, that is, as a varying gear ratio drive.

The mechanism can be used, through suitable addition of ancilliary equipment, to perform differentiation. Murray (1948) explains this. Such a need, however, is less common than that for integration.

The method relies on adequate friction gearing design, the theory of which is laid down in Trylinski (1971). A simpler form, using a disc driving directly onto the 'cylinder', that is reduced to be a translating small drive wheel is also feasible, but in that form the position of the small wheel cannot be changed if the system is at rest without scuffing the drive wheel surface.

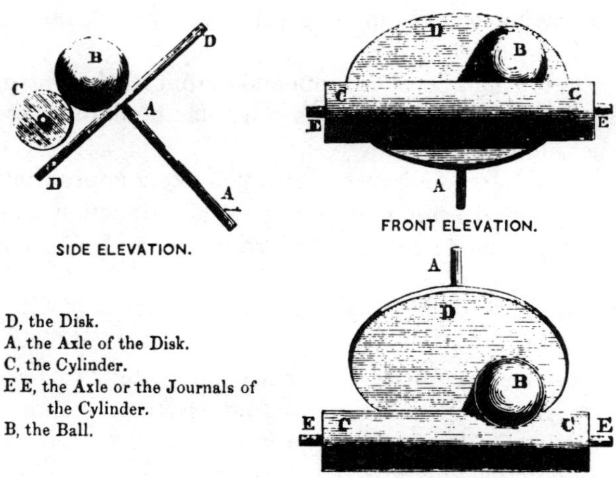

Figure 6.9 Disc, ball and cylinder integrator; Thomson's 19th century etching

It has been shown that several mechanical computing mechanisms are available that might offer an alternative and superior solution to electronic alternatives in certain applications. The references given here are but a little of the extensive literature on the subject of mechanical computation. Further detail will be found in the many reports of actual equipments made from the late 19th century times until the 1950's when the methodology begin to find serious competition from the electrical and electronic alternatives. The account, Eames and Eames (1973), provides numerous illustrations of apparatus. The citation source at the rear of that work enables the original publications to be studied.

References

Beggs, J. S. 1955. *Mechanism*, McGraw-Hill, New York.

Eames, C. and R. 1973. *Computer perspective*, Harvard University Press, Massachusetts.

Mabie, H. H. and Ocvirk, F. W. 1958. *Mechanisms and dynamics of machinery*, Wiley, New York (3rd Edition — 1975).

Murray, F. J. 1948. *The theory of mathematical machines*, Kings Crown, New York.

Pugh, J. 1975. *Calculating machines and instruments*, Science Museum, London, (1st edition — 1926).

Svoboda, A. 1948. *Computing mechanisms and linkages*, McGraw-Hill, New York.

Sydenham, P. H. 1979. *Measuring instruments — tools of knowledge and control*, Peter Peregrinus, London.

Thomson, W. and Tait, P. G. 1883. *Treatise on natural philosophy*, Part 1 Cambridge Univ Press.

Trylinski, W. 1971. *Fine mechanisms and precision instruments — principles of design*, Pergamon, Oxford.

Chapter 7
Limits to perfection — sources of error

The importance of concentrating on error reduction

The earlier chapters have laid the foundations for achieving the mechanism needed in instruments. In each discussion the subject of error has not been covered in any depth as it is more expedient to treat the various sources in a single chapter due to their ubiquitous nature. As each form of error source is introduced, the various approaches for its control and reduction will be explained.

It should be self-evident that any instrument system must finally work as was intended at the design stage and keep on performing to the required specifications. Instruments designed specifically for measurement must also retain input-output relationships in order to provide accurate, meaningful, error-free data. For all of the obvious nature of this statement, it is apparent that many of those who create instruments are not sufficiently aware of the many sources of error that can exist or how to control them. This chapter provides a basis for this important area of instrument design.

Listed are the generally met sources, along with rudimentary explanations, so that they can be allowed for both in the initial design stages and in the practical use of the instrument. Further reading beyond this chapter may be needed.

In many uses, measuring instruments provide the prime basis for assessing if understanding of a subject is correct. It is often very small differences between the current theory of a subject and the real situation that leads to refinement of the theory. This effect was called the 'residual effect' by Herschel. If the errors of an instrument are not adequately known, any difference between measured and expected values cannot be confidently attributed to shortcomings in the theory. For this reason alone errors existing within an instrument and those caused by external influence parameters should be well understood.

In the less scientific use, such as that in general industrial control and measurement applications, residual errors may not be so vital. In such cases, however, other factors require errors to be understood. Examples are the need to be within stated delivery specifications, to operate under reasonable field environments, to perform better than a competitor's product, and to make better use of materials and manufacturing methods. Each of these needs proper understanding and control of errors of design and use.

It appears that no similar general account of errors has been published despite the existence of numerous specific texts on instrumentation. A very few works, mainly dating from many decades before the present, contain useful introductions. These will be referenced as the account proceeds.

Fundamental limits

In any instrument design there exist certain design factors that cannot be eliminated. All physical processes inherently contain physical phenomena that provide a limit to sensitivity. In the majority of mechanical designs this is not likely to be of great significance. (This is, however, far from the case in electronic systems).

In mechanical systems the prime fundamental limit to detection is the Brownian movement, a random motion, of the molecules forming the substance of the parts concerned. The motions of the molecules are averaged by the mass of the part so larger components exhibit less overall dimensional change due to this source than smaller parts. It has been extensively studied in relation to electrical indicating galvanometers in the period before the 1950's and has become of interest again in recent times in relation to the possible detection of gravity wave radiation. Chapter 17 in Bleaney and Bleaney (1957) provides a basic introduction to the theoretical considerations. Braddick (1966) discusses noise sources in experimental setups. Further references are given in Sydenham (1979) where the importance of this limit in the evolution of electromechanical instruments is reviewed.

The bibliography given in Logan (1973) contains references to many of the published papers on gravity wave detectors. Jones (1967, 1968) contains useful practical accounts of the limits to measurement sensitivity.

To provide a rather crude estimate of the level of dimensional noise that Brownian motion effects will generate, some published examples are now quoted. These, however, must be accepted only in their full context as given in the quoted reports. Jones (1967, 1970) estimates that the root-mean-square length change of a 10 mm brass block of 1 mm^2 area is likely to about 10^{-16}m when averaged over a 1 s integration time. This level of change is well above the expected dimensional changes from the weaker gravity waves but in general such extreme detection is not required.

As a second example, Usher (1973), a review of seismometer designs, contains an estimate that a seismometer with a mass of 10 g, a natural period of 1 s, a mechanical quality factor Q of 100, and operated at room temperature will possess a noise equivalent acceleration of 3×10^{-11} g. He shows that the factor for reducing this in that design is the damping, which should be kept low. The natural frequency of the spring-mass system is not a factor in determining the amplitude of the Brownian motion noise. In these instruments this noise source must be considered in order to create satisfactory instruments.

Showing up in yet another form, in a sensitive electromagnetic galvanometer, the source there might give rise to noise equivalent currents of around 10^{-12} A. Further examples are to be found in Barnes and Silverman (1934).

Although not strictly a truly fundamental limit, the effects of creep and hysteresis of mechanical materials often acts as an effective fundamental limit. There are practical levels to which any amount of better design, external influence parameter control, and material processing cannot better. Despite enormous research effort seeking to quantify the fundamental mechanisms of creep and hysteresis in materials there does not yet appear to be available an adequate model of the underlying process. Many texts are available that summarize the current state of knowledge. Design to reduce hysteresis and creep is generally based upon experimental data about the material used and the experience of the designer.

It is suggested that hysteresis can be reduced by limiting the mechanical stress excursion to the smallest allowable cycle amplitudes and by the use of low hysteresis materials (see Chapter 2). Creep is a still more elusive parameter. Clearly releasing locked-in stresses by annealing and other stress relief methods helps. Weathering for prolonged periods is also used when the time allows, for it is known that the creep rate generally reduces with time. A part considered to be creep-free, however, may change its creep characteristics quite markedly if remachined or worked in some way. Sydenham (1973) is concerned with the long-term stability of tensioned Invar and fused silica.

The difficulty faced by the instrument designer is that the vast majority of creep and hysteresis data is compiled for highly stressed parts, that is, those operated at very high temperatures and large stress excursions such as are found in aircraft turbine blades. Instrument designers, on the other hand, are generally interested in comparatively low stresses and well controlled, around ambient temperature. The extreme difference between the two cases makes extrapolation from one zone to the other a very unreliable procedure.

As a guide it is generally good practice to stress materials (those that must be stable) to no more than a tenth of their elastic limit. Brasses are considerably better than steels in as far as creep is concerned in instrument parts. It is not possible to provide any really worthwhile rule of thumb guides for these two parameters.

Both of these factors are given a very poor account in instrument design texts, presumably because of the poor state of general knowledge about them. Trylinski (1971) discusses hysteresis, but not in numerical terms.

Creep is often finally accepted as a factor that cannot be reduced to the levels really desired. Being largely (an often incorrect assumption) systematic and adequately linear with time, it can be allowed for in the measurements made with a drifting device. A prime example is the exploration gravity meter in which an established routine is set up, with respect to time and position of readings, such that the drift can be assessed and removed from the data taken in the field.

Geometrical factors

Attention is now turned to these features of a mechanical design that can be controlled in order to improve the performance and accuracy of the measuring apparatus.

A clearly identifiable group are those sources of error that relate to the geometrical features of the mechanical assembly. These are almost all parameters that can be kept in control by careful design of the assembly and through adequate selection of components, materials and sizes.

Mechanical design of instruments invariably rests upon the assumption that there exist points in the space framework that are invariant in position with respect to each other. They are often merely drawn as cross-hatched lines in schematic diagrams of the system.

In reality all materials are elastic and they all change dimension to some degree. It is, therefore, important to ensure that such elements are, indeed, sufficiently strong, inelastic and stable.

Stresses occurring in manufacture, testing, transport, and assembly are often far in excess of those at the working state. For example, to align a measuring machine on a test bed as the final test of the machine could provide misleading information about the finally installed performance. Bolting down stresses can change the shape; transport might have imposed distortion of the frame.

A second form of geometric error sources is the dimensional relationship change that will occur due to the actual dimensions of individual parts lying in an error band given by the tolerances to which the parts have been made. Tolerances should be selected for the various surfaces of a component according to the assembled geometry required. There is an art in tolerancing; it comes largely with experience. Whereas, to overspecify may well control the assembly dimensions more rigidly, it makes the instrument far more expensive.

Standards, such as BS 1916 on limits and fits for engineering, provide the starting point to the selection of realistic tolerances for given situations. This aspect is generally included in courses on mechanical engineering.

Related to tolerance error is the gradual loss of accuracy that might occur if a wear-true philosophy (see Chapter 3) has not been adopted. Wear in pivot points and along slides will change the effective geometry of the instrument, thereby possibly degrading the performance.

Such matters, plus other general aspects of design, are gained largely by direct experience. Several authors in the past have endeavored to pass on that experience as descriptive guides to general instrument design. These include Braddick (1966), Glazebrook (1923), Rolt (1929), and Whitehead (1954).

Rather more specific are errors caused by assuming that location points, such as a center or a sliding perpendicular, are provided in a perfect manner by a bearing or a slide, respectively.

Abbe's Principle

Abbe's Principle, although well known among practitioners, does not seem to have been defined in English language texts about instrument design. Basically, it states that a measurement is generally best performed by applying the measuring device to the object in as direct a manner as possible. For example, in the field that this was first expounded — mechanical dimension measuring equipment — it is preferable to determine the length of a part using a comparison method that introduces as little error due to the comparison process as is practicable. Figure 7.1 contrasts the use of a scale to measure the length of a part with the method in which a travelling microscope head slides along a scale. Where the scale markings and the interval that must be measured can be placed in actual contact, the reading cannot include any error from the comparison process. (Of course, it could contain error of the scale or from its misreading.) In contrast, the travelling microscope arrangement can introduce error by way of the slide not being straight, causing the microscope to project the scale incorrectly on to the unknown.

Whereas Abbe's Principle might be the ideal to aim for, it is often not practicable to adhere to it. If not, then other geometrical errors must be considered.

Sine and Cosine Error

Angular variations introduced by the mechanism embodied in the travelling microscope example given below may, or may not be, serious. It depends upon the effect on the length being determined or required to be held constant.

If the same microscope is used at each end of the length and the slide is straight (Figure 7.2a), any fixed angular error, α, in the mounting of the microscope head will not cause error in the length determination because it is equal at each end. If, however, the microscope mount is perpendicular to the slide at one end and the slide causes the head to rotate by angle, α, at the other end (Figure 7.2b), an error equal to $d \sin \alpha$ will arise. As this error is directly additative and because sines of small angles are significant quantities in such cases, quite small angular changes can give rise to relatively important error magnitudes.

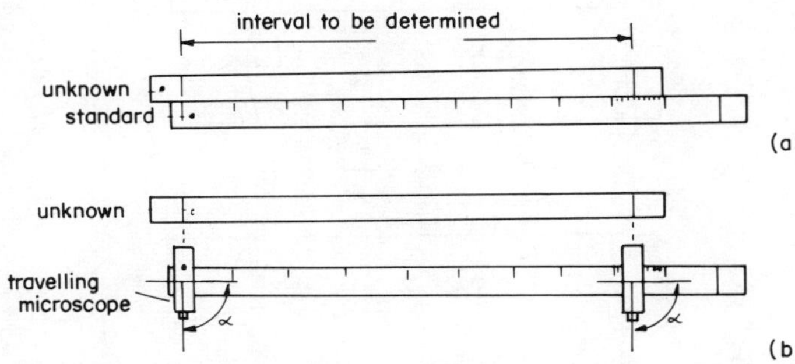

Figure 7.1 Abbe's principle. In (a) the unknown and the standard are compared in a direct manner but in (b) errors may exist in the projections of the scale by the microscope.

When measuring a length (Figure 7.2c) between two end intervals instead, as arises with an inside screw micrometer (the above sine error case occurs in vernier calipers), then a second kind of error source can occur. If the part to be measured is not aligned with the central axis, the length will be $l \cos \alpha$, not l as indicated. This is not as significant as the sine error of the previous case for if the angle α is small, which it usually is, the $l \cos \alpha$ is closely the same as α. Another example where this error arises is in the setting up of a distance measuring ranger. The actual alignment need not be very accurate, because the error caused by misalignment will not be significant.

Where cosine errors are generally easy to control, they must, nevertheless, be calculated to ensure that the approximation holds to the degree of accuracy that is required. Sine errors, similarly, must be controlled by ensuring that the angles remain within tolerable limits under practical use. As d is reduced, in Figure 7.2b, it will be seen that Abbe's Principle is approached.

Errors of this form are discussed but only briefly under the principle of alignment in Rolt (1929) and in Braddick (1966).

Centering Error

Where the measurement of location depends upon a bearing around which motion takes place, there may arise errors of centering. It is common

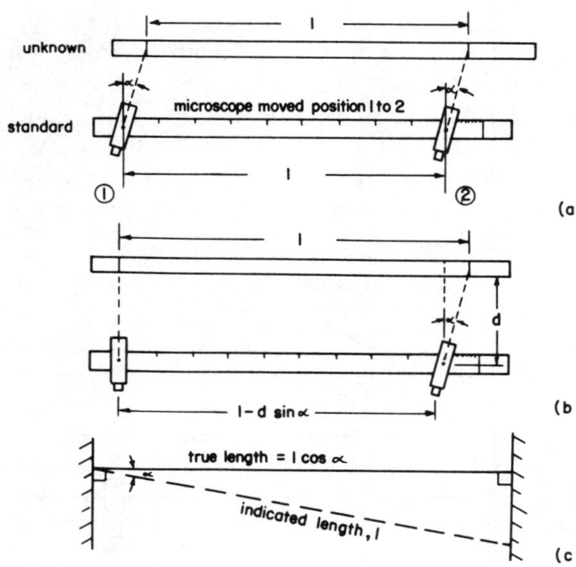

Figure 7.2 Sine and cosine errors. (a) error of microscope mounting not significant (b) sine error occurs at one end due to rotation of microscope (c) cosine error

knowledge that such errors can be reduced by reading the scale on opposite sides, taking the average of these as the correct value. Whitehead (1954) devotes space to this problem, pointing out that it is not always necessary to use the best possible bearing tolerances. He provided an analytical study of the magnitude of errors. Many instruments are made to be read at, at least, diametrically opposite places. Indeed, some arrange more than this number of averaging processes to be carried out in provision of the output value. Whitehead, however, only considered the use of a centering bearing for a radial scale. In linkages the error effect must be considered from other viewpoints. Braddick (1966) also provides material on centering errors.

Although the above error sources have been illustrated here in terms of the traditional, manually read instruments, it will be found that these error sources arise frequently in the design of the mechanical features existing in transducers of the modern form.

Influence parameter errors

The very best of internal designs can only provide satisfactory operation if variations in external influences either lie within tolerable limits or can be handled by the internal design arrangement.

All systems are influenced by changes in ambient conditions to some extent. It is the relative magnitude of these effects that matters. Temperature, ambient air pressure, relative humidity, magnetic and electric fields, acoustic noise level, mechanical vibration, chemical attack, and more can each possibly give rise to error in operation of the installed instrument.

It is common and established practice to test an instrument toward the end of the design cycle for its tolerance to such effects. Calibration is often accomplished taking into account given levels of vibration of influence parameters. It is, however, often overlooked that the test conditions may not be reproduced in the final use situation where yet more influence parameters may be in existence.

The designer of an instrument must know which of these influence quantities will exist in use and how to allow for them. There are two approaches. One is to control the internal environment of the instrument by the use of a suitable enclosure; for example, the internal evacuation in a vacuum weighing balance that avoids errors of buoyancy and provides some degree of thermal isolation. The alternative is to design each element of the system such that the overall system can tolerate significant enough influence parameter variations. An example here would be to use low thermal expansion materials plus certain mechanical structural arrangements (that will be discussed below) to reduce thermal differential expansion error to within tolerable limits for the expected design temperature variations.

In practice, one or the other or a combination of these two alternatives is used. Probably the hardest part of the task is properly and adequately to decide what influence quantities are relevant and their magnitudes of change.

Temperature Error in General

Without doubt the most prevalent influence parameter in the mechanical design of instruments is temperature change. It has been shown in Chapter 2, Figure 2.2, that a material with absolutely zero thermal coefficient expansion does not exist. Furthermore, materials with the lowest values are not always suitable for the manufacture of instrument parts. It is, however, possible to produce instruments from the relatively high coefficient materials, which are reasonably temperature insensitive. Before going into the methods that might be used, it is instructive first to examine the general thermal design of a mechanical instrument.

Thermally controlled jacketing and mechanical compensatory mechanisms often are used to overcome temperature error. The value of each generally rests on an underlying assumption that within the instrument the temperature is uniform. This is often far from the case. It is not adequately recognized that thermal heat flow is not an instantaneous process. Any change of forcing temperature on an instrument will give rise to changes in temperature of the various parts at varying rates. If, in exacting designs, the instrument is regarded as being a thermally dynamic rather than static assembly, then better design should result.

If the external temperature does not change, then only internal sources of heat such as arise from electronic circuitry enclosed in the housing or from friction between moving parts can cause thermally produced differential dimensional errors. Heat source changes due to adiabatic action on air pressure may be relevant. Design, therefore, should aim to keep internal and external sources of temperature change as small as necessary. (They cannot ever be held at an absolute stability.) Internal heat transfer mechanisms should be arranged so that temperature uniformly results as speedily as is required. In practice these simple rules are very difficult to adhere to.

Consider first the problem of thermal uniformity. The materials that are most suited for construction, such as the steels, aluminiums, and brasses, do not have very high thermal conductivities. They also possess comparatively high thermal expansion coefficients. Lower coefficient materials generally have still lower thermal conductivities.

Heat is transferred by conduction, convection, and radiation. Knowing the constraints imposed by the materials that have to be used — good thermal conductors are not the best of structural materials — it is possible to arrange for heat transfer to be improved through the use of design layout, shape, cross-sectional size, and similar design possibilities.

As a guide, design for good thermal uniformity will aim to provide good conduction inside the instrument with a high thermal barrier attenuating external changes through the use of some form of integrating enclosure. Enclosure design can also be most sophisticated. Reference will be made to that later.

To obtain good internal uniformity, all parts should be made to maximize conduction through the use of high thermal conductivity materials, large cross sections, good thermally conducting joints, and optimal arrangement to keep conduction paths as short as possible.

Conduction and convection can often be improved by the use of a stirring arrangement such as a fan or stirrer paddle. The internal fluid, in such cases, will be a suitable liquid or gas that will assist heat transfer. The transfer of thermal energy is, however, often decided more by the heat transfer film coefficient existing at the interfaces than it is by the fluid's internal properties.

Sometimes the design aims to prevent convection taking place, thereby aiming to maintain the temperature constant, if not uniform, within the instrument. It is a physical fact that the smaller the convecting cell size, the lower the rate of convection. Below a certain cell size and shape it ceases completely. Because of this, it is often helpful to fill interspaces of an instrument with loose thermal insulation so that the internal convecting spaces are made smaller.

Heat transfer by radiation can be increased by suitable use of surface treatment, blackening being the most used. The actual surface treatment used will be decided by the thermal wavelengths involved. Within the system, parts are blackened and the whole is then surrounded by a thermally reflecting enclosure. The silvered vacuum flask is a good example of this.

It will be appreciated from this account of the possibilities that the thermal design of an instrument can be very complex. Perhaps that is why this aspect of design of instruments has not been adequately researched or organized. Braddick (1966) gives brief mention to the problem. Considerable practical information exists in primary journal literature on temperature controllers but the basic principles, theory and design do not appear to have yet been ordered in a didactic manner.

Basic texts on heat transfer, such as Eckert (1950) and Jakob and Hawkins (1957), provide general design principles. Publications relating to cryostat design (White, 1959) and use also delve in these problems. Abachi *et al.* (1979) provide a recent bibliography of reported designs. Designers of microbalances have given thermal effects considerable attention, Wolsky and Zdanuk (1980) being a key review work. Walsky *et al.* (1967) discusses how gold plating a quartz balance arm can reduce thermal expansion errors through subtle heat transfer mechanisms. Sporton (1972) is probably the best available paper to consult when beginning a design of an enclosure.

Thermal enclosure design is often needed in the development of the total instrument system. Some applications require the instrument to operate within temperature limits as small as a mK variation. Design aims to form an enclosure in which the walls seen by the contained instrument are uniform in temperature and of the lowest practicable thermal impedance.

Simpler designs generally use an almost point source of heat, which is controlled from a single point on–off temperature sensor. Clearly this does not come close to the rule of maintaining uniformity throughout and, therefore, only provides crude control. Far better are the designs in which the heater and the sensor are distributed throughout the wall area of the enclosure. Proportional control is generally assumed to provide the best control loop but this may not always be the case.

Parameters of thermal enclosure design are the external variations and frequency to be controlled, the size and shape of the instrument (clearly smaller instruments are easier to control to a given limit than large structures), the limits of control variation required, the internal heat sources, the enclosure design (sandwiched heat conducting and insulating walls), plus judicious use of air spaces have proved very effective (Sporton 1972).

In many applications the aim is to integrate the external thermal forcing function to an adequately small internal variation. Thus it seems an obvious step to use an enclosure that has an adequately long thermal time constant. In practice this can be, with passive methods at least, less effective than might be expected. The thermal time constant (that for a 64% change if the system is assumed to approximate the first-order linear response) for a vacuum flask is around two to three hours, a solid house brick box is similar. When using such methods to average out daily variations, they are barely effective. Daily variation smoothing requires several meters of rock cover.

Active systems, those in which temperature is actively controlled, can be far more effective. Long time constant passive surrounds also suffer in that they must be given time to stabilize before the internal instrument can be used in its intended state. As thermal mass is the dual of thermal insulation it is possible to construct first-order thermal enclosures from either conducting or insulating material. Each has its advantages. Use of both kinds together raises the order of the enclosure's thermal time response and, therefore, if designed properly, can provide better results.

In the practical construction of thermal controllers, it is far easier to produce heat energy than it is to remove it (compare electric heating elements with Peltier cooling cells, for instance). For this reason thermal surrounds often are operated at some margin above the ambient temperature so that natural heat loss acts to provide the heat reduction mechanism.

It is the writer's experience that a contributing factor to the mechanical instability of materials is the stress brought about by temperature variations of the part. It appears that the more controlled temperature variations are, the better the ultimate stability of creep exhibited. This is further supported by the practice of thermostating gravity meters with a temperature-controlled jacket that is kept energized even when the instrument is not being used.

The alternative method for reduction of thermal mechanical error is the use of compensatory mechanisms.

Most constructional materials possess a positive coefficient of thermal expansion; that is, they expand in length with increase in temperature. Due to the shortage of suitable, large enough, negative coefficient materials (they do exist — carbon fiber and the silicas), it is not generally possible to counteract the positive expansion with a negative part as shown in Figure 7.3. This kind of component, however, using two different positive coefficient materials, can provide an adjustable positive coefficient rod by clamping it with distance *d* spanning the ratio of the two lengths required.

This problem can be completely overcome (within limits) by the use of two positive coefficient materials arranged as a re-entrant compound structure. Figure 7.4 shows the mounting arrangement used to maintain the reference arm distance, *d*, in a Michelson optical interferometer constant with temperature change to the mount and corner cube. As the cast iron expands outward with temperature rise, the brass expands in the reverse direction at a larger rate, giving a net change of zero to distance *d*. Concentric tubes can also be used to form a zero length change bar. Bimetal strips, in which the two materials are joined, can also be used to change dimension in some way through a forcing action.

The choice of materials being limited means that the design of re-entrant arrangements must proportion the relative lengths to obtain the zero (or other) state using the small range of thermal expansion values that will be found to be usable. In the above mount the design also allowed for the expansion of the glass of the corner cube.

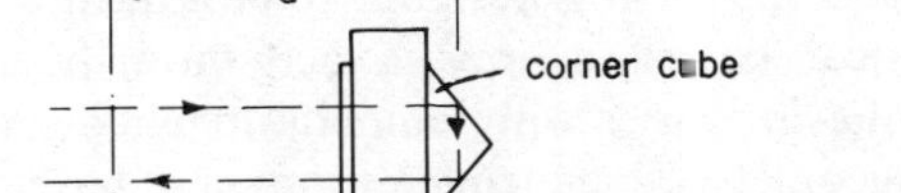

Figure 7.3 Thermal compensation by cascaded bars

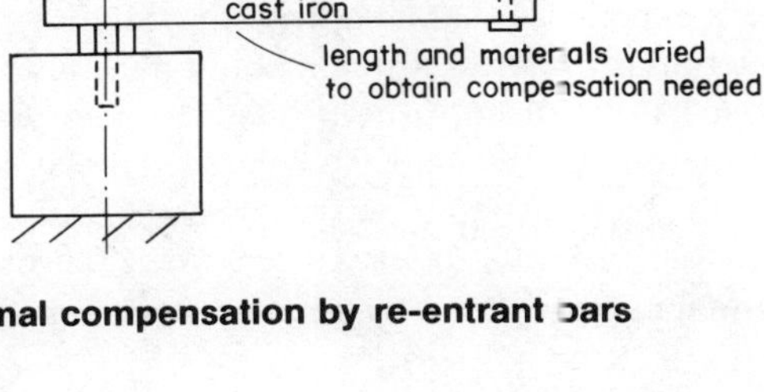

Figure 7.4 Thermal compensation by re-entrant bars

The assumption that both parts of a re-entrant link change (track) temperature together may not hold; during temperature transients the compensation may not operate satisfactorily.

A second method, one that is not used so commonly as the re-entrant arrangements (because, it seems, it is not well publicized) is to make use of the stresses set up by temperature change in an element. Baird (1968) used this principle (Figure 7.5), to compensate a bar of Invar.

In his method steel wire was wrapped around an Invar rod. It was then soft-soldered to it, so that axial stresses in the steel wire, caused by temperature changes, are coupled into the circumference of the Invar bar. With temperature rise, the steel attempts to form a larger diameter ring around the Invar, causing it to be tensioned outward. This, in turn, via the Poisson ratio shape-volume change effect, causes the Invar bar to reduce in length. The reverse holds true. He provided the theoretical considerations for obtaining the degree of compensation needed for a pair of materials. Clearly this method can be applied by plating layers, or by pressing tubes, over the base bar. The thickness and the external material used are the key design factors.

A third method for compensating thermal errors is to measure the temperature change in a part, applying a generated appropriate length change where needed. Many methods exist for the provision of small length changes under electric control. They include piezoelectric crystals, magnetostrictive rods, and thermal heating of rods and electromechanical devices. Measurement of the temperature of a part can be difficult because of transient thermal effects. The control loop may also introduce phase shifts.

Temperature Error of Elastic Elements

Temperature changes to elastic elements will give rise to length changes according to the thermal coefficient of length; additionally, however, the elastic modulus will change.

The combined effect is to alter the elastic rate (scale error) of the spring element and also to alter its free length, giving rise to a zero position shift (zero error). If the component is designed to be a flexing spring, then these combined effects can be very serious. Exactly how the spring's combined scale and zero behavior varies with temperature change depends upon the spring form and geometry. Producing a spring that has negligible temperature error does not occur merely be selecting a zero elasticity modulus material, not will it occur by using a zero thermal coefficient of expansion material.

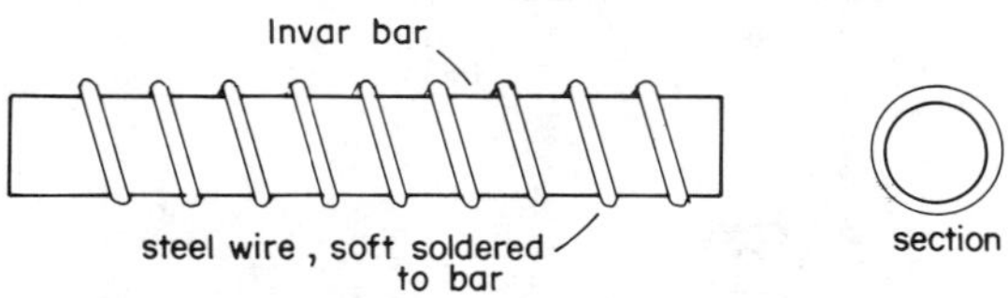

Figure 7.5 Thermal compensation by counter-stressing

It can be shown (see Gitlin, 1955a) that for the common cantilever, helical, and torsional forms of spring the error of spring rate change (scale error) is not a factor of the physical dimensions but it is dependent upon the thermoelastic and linear expansion coefficients and the magnitude of the temperature excursion. Thus, once the material is chosen, a given level of scale error exists for all designs with that material. Beryllium copper, for example, possesses a scale error of –0.03%.° / C for an increase in temperature. That is, the spring softens with temperature rise.

Shift in zero with temperature is dependent upon the spring's form. The three articles by Gitlin (1955a, b, c) are key reference accounts about the thermal errors in instruments. Trylinski (1971) includes discussion of this problem.

Error Due to Ambient Pressure

The mechanical components of an instrument change dimension according to the external pressure exerted upon them by their ambient. This has a very small effect, but on occasion it can be significant. It is calculated from the Poisson ratio for the material, which expresses the longitudinal change in length for transverse stress variation. As a guide, metals will lengthen by around 1 part in 10^8 for a typical daily ambient barometric pressure variation.

Error from this source can be reduced by the use of ambient pressure control, which need not necessarily be at the evacuated level as long as the pressure remains constant enough. Alternatively, it can be compensated in some way through the use of some form of pressure-sensitive device.

Sealing an instrument inside a container does not necessarily provide constant pressure inside the container; temperature changes to the gas inside it will give rise to significant pressure changes.

Miscellaneous Error Sources

The above discussion has dealt with sources of error occurring in mechanical components. Also of importance might be those cross-coupled, mechanically produced effects that transduce mechanical energy into another energy domain. An example is thermoelectrical current, which is generated by the mechanical materials used in combination with thermal effects. Chapter 2 lists many of these. Piezoelectric and magnetostriction are electrical resistance-strain changes that are each brought about by mechanical input parameters. They must, therefore, be considered in the mechanical design stages of an instrument as possible error sources in the total system.

It can be seen that there are many sources of error in, or caused by, the mechanical components of an instrument. If good performance is to be obtained, these must be considered and reduced to satisfactory levels of magnitude. This point needs stressing because this aspect of mechanical instrument design is sadly neglected in texts, in courses, and in practice.

Scientists involved in the manufacture of their own experimental and measurement apparatus often reach these barriers and they therefore are aware of the errors that can arise. Other designers, however, often do not realize that their designs also may suffer from such effects; it is necessary to eliminate them as design parameters by reasonably rigorous study before they can be honestly regarded as insignificant. They are often easily reduced at the initial design stage.

References

Abachi, H., Molenat, J., and Malbrunot, P. 1979. 'Continuously variable cryostat giving accurate settings at any temperature between 77 and 300K'. *J Phys E: Sci Instrum*, **12,** 706–711.

Baird, K. M. 1968. 'Compensation for linear thermal expansion'. *Metrologia*, **4,** 145–6.

Barnes, R. B., and Silverman, S. 1934. 'Brownian motion as a natural limit to all measuring process'. *Rev. Mod. Phys*, **5,** 162–192.

Bleaney, B. I. and Bleaney, B. (1957). *Electricity and magnetism*. Clarendon Press, Oxford.

Braddick, H. J. J. 1966. *The physics of experimental method*, Chapman and Hall, London. (Edn. 1 — 1954).

Eckert, E. R. G. 1950. *Introduction to the transfer of heat and mass*. McGraw-Hill, New York.

Gitlin, R. 1955a. 'How temperature effects instrument accuracy', *Control Engng*, April, 70–78.

Gitlin, R. 1955b. 'Whats available for compensating instruments for temperature changes'. *Control Engng*, May, 70–77.

Gitlin, R. 1955c, 'How temperature compensation can be used'. *Control Engng*, June, 71–75.

Glazebrook, R. 1923. *A dictionary of applied physics*. Vol III, 445–450.

Jakob, M. and Hawkins, G. A. 1957. *Elements of heat transfer*. Wiley, New York.

Jones, R. V. 1967. 'The measurement and control of small displacements'. *Physics Bull*, **18,** 325–36.

Jones, R. V. 1968. 'More and more about less and less'. *Pro R Instn Gt Br*, **43,** 323–345.

Jones, R. V. 1970. 'The pursuit of measurement'. *Proc IEE*, **117,** 1185–1191.

Logan, J. L. 1973. 'Gravitational waves — a progress report', *Physics Today*, **26**(3), 44–52.

Rolt, F. H. 1929. *Gauges and fine measurements*. Vol II, chapter XIV., Macmillan, London.

Sporton, T. M. 1972. 'The design of a general-purpose air thermostat.' *J Phys E: Sci Instrum*, **5,** 317–321.

Sydenham, P. H. 1973. 'Nanometre stability of Invar and quartz suspended in catenary'. *J Phys E: Sci Instrum*, **6,** 572–576.

Sydenham, P. J. 1979. *Measuring instruments, tools of knowledge and control*. Peter Peregrinus, London.

Trylinski, W. 1979. *Fine mechanisms and precision instruments — principles of design*. Pergamon, Oxford (Translated from Polish by A. Voellnagel).

Usher, M. J. 1973. 'Development in seismometry'. *J Phys E: Sci Instrum*, **6,** 501–507.

White, G. K. 1959. *Experimental techniques in low-temperature physics*. Oxford University Press, Oxford.

Whitehead, T. N. 1954. *The design and use of instruments and accurate mechanism*. Dover, New York (Edn 1 — 1933).

Wolsky, S. P., Zdanuk, E. J., Massen, C. H., and Poulis, A. 1967. (On gold coated quartz arms to reduce thermal expansion). *Vacuum Microbalance Tech*, **6,** 37.

Wolsky, S. P., and Zdanuk, E. J. 1980. *Ultra micro-weight determinations in controlled environments*. Wiley, New York.

Chapter 8
Realization of prototypes

General comment

This final chapter provides, through experience with the design and development of an electromechanical instrument, an account of the kind of procedure used to produce a well designed prototype instrument. It does not attempt to discuss how that model is converted into a production instrument — that stage is concerned with other specializations that go beyond the aim of this text.

Although now rather dated in that the technology is restricted, the description of design given in Whitehead (1934) is possibly of value to the new designer as a supplement to that presented here.

In Chapter 3 some generalities of design of mechanical systems were introduced. That aspect is first taken up and expanded before considering a specific case.

The essential ingredients that go into the creation of a new or improved instrument are people — their attitudes and skills; materials — their availability and suitability; theoretical principles; processes and tools of manufacture; and the institutional mechanisms that foster instrument design in some way.

At any given time the designer can only do as well as the state of these factors will allow. Even today, after at least 6000 years of instrument development, we seriously lack many materials, our manufacturing processes can still be bettered, there are new principles being discovered all the time, and there is certainly a need for improvement in the way instruments are generally designed.

Design of the measuring apparatus begins using the more obvious ways to proceed, gradually extending detail and becoming more complex as the important factors are identified and detailed. Intuition has always played a large part in design methodology, and it still seems that a design begins with the designer imagining certain structures and systems organizations that are then tried out and systematically improved until they come up to the specifications thought to be satisfactory. Figure 8.1 illustrates how the designer gradually progresses from the general to the specific, mustering the limited resources to create a satisfactory system within time and cost constraints.

Although this chapter concentrates on prototypes, (which generally means one-offs) it must be stressed that all design should give recognition

(Note: the case study part of this chapter was compiled in close cooperation with Dr. A. L. Hugill, then of School of Earth Sciences, Flinders University, Adelaide, South Australia, now with Scintrex Inc., Ontario, Canada).

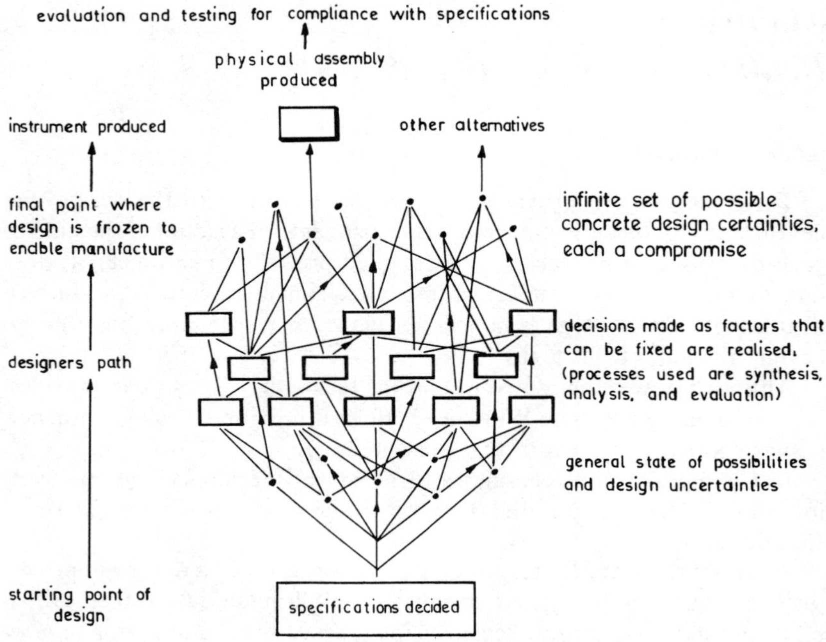

Figure 8.1 One representation of the design process

to procedures that would allow the prototype to go into the production phase if that were forthcoming. Too many first designs have been restricted from further development by the use of practices that have later meant considerable rework when further identical units are needed. For example, the use of obsolete components or parts made that are not conventional would result in further editions needing redesign to accommodate replacements for the impossible-to-obtain parts.

Good prototype design, therefore, also requires of the designer some experience with what would happen to the initial design if it went into production. An impression of the production team's problems can be gained from study of Figure 8.2 where the stages needed to bring an idea into a manufactured, marketed product are shown. In terms of time taken making a prototype may, at the time, seem a very long task, but compared with the time to take it into production, it is small. As an example, to market a relatively simple mechanism might need as much as two years to pass from the decision to market a specific product to sales being ready to proceed. Designing and building the first pre-production instrument may take only a month or two! This point is not appreciated by many instrument designers, especially those who make their own instruments for a scientific kind of task without production being, at the time, of any interest. If persons in this situation were more conversant with the developer's prob-

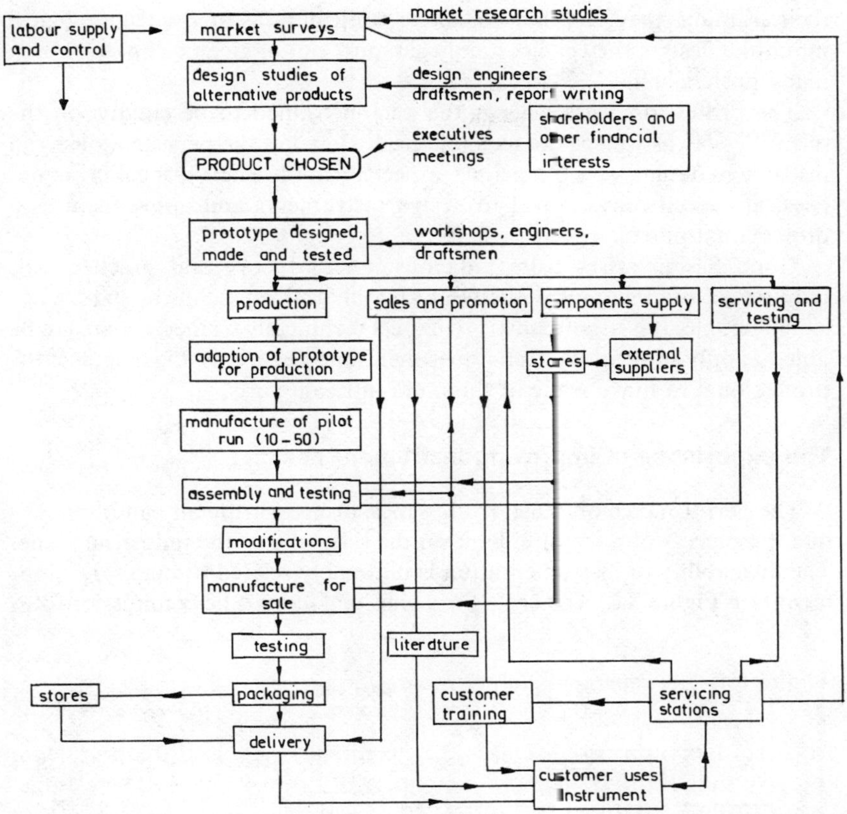

Figure 8.2 How an instrument develops within commercially based manufacture

lems, they should be able, with little inconvenience, to produce the instrument they need which is also more suitable for mass production.

Instruments are still largely being made by methods that were in vogue a long time ago. Production batches are often as small as tens of articles. A few hundred identical instruments in one batch is regarded as a large size. For this reason, the use, where possible, of already available, mass produced subassemblies is preferred to keep costs down and design flexibility up. Unless very large numbers are going to be made it will not be practical to consider the use of mass production manufacturing methods such as die casting or automatic manufacture and assembly (exceptions exist).

For too long it has been generally regarded that instrument design, manufacture, and application are more in the realm of the technician than the professionally trained engineer or scientist. Technicians are trained and skilled at applying known technology. They are, wrongly, often expected to advance the state of the art with new scientifically based ideas. By reason of their training, they can, indeed, make contributions at the technological input to a design; they can certainly get more out of given technology than many professionals.

The professional graduate is the person trained to be creative in the scientific way as well as the technological. Here lies the persistent defect in instrument design; very few people, especially in the English-speaking countries, are specifically trained to design instruments and apply them as a professional pursuit.

Good design arises from judicious use of theory and practice; this requires a close relationship between suitably trained technical and professional people. It is as unrealistic to expect technically trained persons to be able to apply some of the techniques mentioned below as it is to expect the professional to make some of the more intricate parts.

The importance of improving instrument design

The performance obtained from a measuring instrument can often dictate the success of a venture, be it on the scientific or the industrial scene. The availability of instrumentation is intimately related to man's development (see Figure 8.3) The costs for a ship that cannot be commissioned in

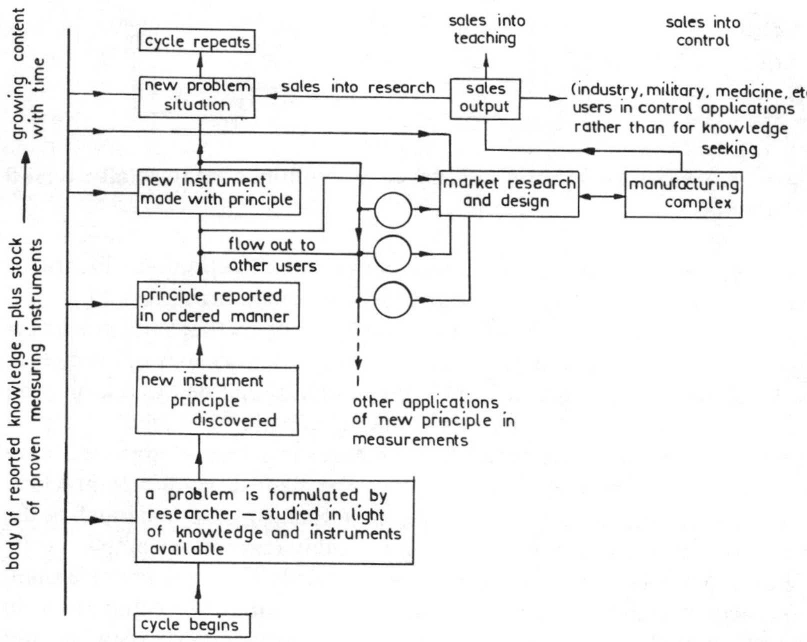

Figure 8.3 Instruments evolve through scientific knowledge passing into new instruments forms which have marketability

time or the plant that fails during operation run into vast sums of money per day. In science it is the instrument performing in a way that assures its user that the discrepancy resulting between the known theory and the actual measurement data is in the theory that lays open the way to new knowledge. Herschel, around 1800, appropriately called this the residual effect. Errors of an instrument, and they all have them, must be clearly known if the user intends to make use of the data in any exacting way.

The apparent simplicity of the generally used design method for instrument development — that process of trial and error that can be used in these cases (compare with using the same method to design a 500 MW turbine!) can lead one to think there is little need for a better method of design. Although this is the way that much of the design of instruments is pursued, it is not to be recommended.

The ad hoc approach does not lend itself to discovery of the means to make radically different and valuable improvements. Mathematical modelling of instruments, which is still in its infancy, can give the designer the chance to try various strategies of improvement that are based upon sound scientific understanding. These can be easily tried out by altering the computer program of settings of the medium in which the model is being simulated. Tailoring the design through a string of prototypes is largely avoided — design features are simulated in the model.

Efficiency in design leads to efficiency in other areas. Freed from a dominant craft form of realization, the designer is able to concentrate on other aspects, such as sensitivity to influencing parameters, cost, serviceability, and the like. Many of these factors are more readily studied if the design follows systematic procedures and uses a more solid foundation of theory.

Widespread systematic instrument design leads to standardization of design procedures. Solid-state integrated circuit (IC) electronic units are, for example, attracting users into choosing commercially available systems blocks rather than calling up costly specials.

Systematic design of instruments shows where the shortcomings exist in current procedures. Much of the present time consuming procedures (such as working out the numerical values of an equation) can be left to machine processing. This frees the designer to concentrate on the important choices that occur each time a branch point is reached, points at which the designer must choose between many possible options that seem to be suitable at the stage of the design.

Consider, for instance, materials selection. Without difficulty it is possible to list over 60 parameters that should be considered when selecting a material to make a part. As going through such a list by manual means is prohibitive, we usually rely on our sixth sense to know what is the right choice. A computer-backed system could presort these, presenting a short list to the user who would be asked to answer the right kind of questions. This method of prompting would help to prevent the designer ignoring important decisions.

Systematic design enables the designer to make use of the powerful computer-based aids now available to remove the routine operations of design. Complex decisions can be made easier.

The final reason that will be given here for improving instrument design is related to dealing with the more extensive measurement situation — the aircraft instrumentation system, the hospital intensive-care unit, the process plant. Designing just one single stand-alone instrument can be complex enough; coping with dozens of them when they are each interacting with each other, requiring different schedules, different calibration intervals and more, needs systematic design control procedures. The problem rapidly grows beyond the scope of the designer's shorter-term memory capacity. Means are needed by which the factors can all be kept in control. This area has steadily risen in importance as systems grow larger, more complex, and more sophisticated. Methodology is emerging for this situation, but such method relies on internal control of the design of its various parameters. Standards of practice, terminology, physical units, and specifications are all needed. They emerge by systematically improving the design of extensive instrument systems, thus resting upon the level of design deployed at the subsystem level.

Summing up current instrument design

In order to show how good design proceeds, it is instructive first to summarize how it is generally practiced. Current instrument design often lacks an orderly procedure that is based upon a systematic use of fundamental knowledge. Design proceeds in several different ways; consider the method for scientific and commercial applications.

In scientific pursuits instruments often develop from ideas presented in published papers and application notes, the necessary modifications being made as they are seen to be needed. Often an instrument system is built using proprietary general purpose units assembled to form the required system function by appropriate settings of the multifunction possibilities. Tests are used to show up any defects in the system and modifications are made to rectify error sources. Seldom is a very thorough study made of the instrument needs at the start, a dominant reason being that the user in many scientific applications often has little information available at that stage on which to make the best decisions. The act of measuring considerably helps tune the user to the problem. As the scientist will seldom have specialized in instrument systems, success will generally only follow after the user-builder has walked the same path as many before him.

Instruments built by this trial and error procedure are expensive (if all of the costs are included), often still far from properly designed, often incorporate out-of-date components, often could not easily be copied or exploited because of poor records and the use of obsolete components. Furthermore, they too often go to press in journals wherein the vital working detail is deliberately left out of the description because of editorial policy that dictates that journals report only fundamentals.

The picture portrayed here does not, of course, represent all of instrument design in science but would apply to the majority of smaller groups. Those groups that have the services of specialized design teams and instrument kinds of workshops will naturally do considerably better in the execution of the task, but they, too, will often lack an orderly procedure of design based on fundamentals. In many cases the actual details, which can be very important in instrument design, are left to the draftsmen to fill in.

In industry the scene is similar in that there, too, fundamental processes of design are often missing in the R & D laboratory. If the firm is making instruments to market, they will, after deciding what to market, usually purchase a representative range of their competitors' products. These are dismantled, tested, and perhaps modified. These instruments provide the modelling facility that helps the designers see what are beneficial features for that kind of instrument.

Finally, the new design is developed. It then looks very tidy in its appearance and will seem as though it was very systematically generated. The chances are that the designer's experience has provided the bulk of the design in which a more technical approach than a systematic scientific one was adopted. As departures from the desired specifications are highlighted by tests, they are rectified. In many cases they are, unfortunately, not eliminated by better design but are cancelled out by a further addition of circuits or parts as best as an add-on modification can yield.

In some cases the instrument marketing firm is not beginning a new design but improving an existing line. This may arise as the result of a good idea or as the result of the sales surveys and servicing records. These modifications are generally not achieved by the use of in-depth scientific understanding but by experimental procedures for the simple reason that too few easy-to-apply scientific procedures exist. Experimental development is often far faster and cheaper than the mathematical scientific way when a short-term answer is required. It, however, lacks substance for future expansion on a solid fundamental base.

If the above gives an impression of disorderly design being commonplace in instrument development, then the desired message has been communicated. It is that "easier to criticize than be correct" is an apt adage for the prevailing situation, for all the best intent will not make much change to this procedure in the short term. Current methods are used because they get results with no other reasonably low-cost alternatives being offered. This situation exists because there has been too little effort devoted by that section of the society that has the opportunity to help — the academics and certain government staff. It is unrealistic to expect industry to seek more streamlined instrument design procedures when instruments are but a little part of the bulk of their design problems. Furthermore, as all other marketing firms are in the same situation, there does not seem to be an anomaly. It is also not likely that an instrument design firm that decides to go 'systematic' will reap the full benefit without it flowing on to competitors.

Future expectations — some procedures to aim for

Recently, however, several academic groups have been concentrating on fields of endeavor that will help in this regard. After many centuries it does seem that we may soon have available systematic procedures for designing and evaluating instruments and instrument systems.

A first difficulty met is often to obtain realistic and adequate specifications of the desired performance of the instrument. In scientific use it may be acceptable to develop the specification as the design grows through use. In other cases, such as the development of a new line for sale or when writing the letters to go out for bidders, the specifications must be decided virtually once and for all, before work begins. This most important stage seems to be the most poorly defined part of an instrument design. What is needed is some method that helps the specification writer recognize the parameters that are important. Computer-aided 'aide-mêmoires' seem to be the answer.

Current texts on instruments often contain lists of what to think about in a design. They usually overlap to some degree but never take in all that is needed. A computerized memory could hold the main options, the memory being added to as further parameters are realized. In this way a specification study of a new problem could then be done in conjugation with the 'computer prompt' gradually developing the key specifications, and what one needs to know about them, in order to make a good definition. Strategies are needed to weed out any content that overlaps. All processes of a design that *can* by done by the machine *should* be done by it; this will enable the designer to proceed with design as fast as possible, using his unique skills of design to the fullest.

The same techniques could be used to organize the design procedure. Design of any product function or the like is basically a process of massive iteration, using guided trial and error, of a many-parameter problem in which few parameters have exact enough knowledge established about them. The computer could guide the designer along each necessary path. When that line is completed or exhausted as unproductive, it would then take the user back to where he or she was before, even if the work were done months previously. Prompts could be issued to refresh the user's memory about the state of the design.

Overall control of the many branching procedure would be needed to provide control on speediness, cost, time utility, and the like. Several designers could be working into the program together. The whole design task might take a considerable time — this is not a suggestion for a quick solution on a terminal but a means for retaining better long-term memory for the designer who is apt to forget past experience or become tired with progress.

Where the design called for specialized techniques (for instance, the design of linkages for given motions), the machine would provide a quick primer about what procedures it has available, given the fundamental issues involved. Then the machine would carry out the many very tedious calculations, displaying only the results for approval by the designer.

An advantage of adopting machine-based design aids is that a rigorous systematic procedure soon highlights the nonscientific and hazy aspects of design. It clearly identifies which areas of the general design of instruments are weak. Further research could then concentrate on improving these areas.

The scenario can be extended far beyond this. Imagine being plugged into a very 'intelligent' terminal that, when data are needed on the properties of suitable material for a component, it merely decides what to search for and where, coming back to the user with the necessary information for final selection and approval by the designer. Imagine knowing that the procedure being used is the latest that has been developed and that information presented is not likely to be much behind the best existing. Such progams do exist in many fields of design.

Such a vision may seem to be far-fetched but the reality is that instrument design is not really so complex and unmanageable as it might appear. It is far easier to list the principles involved that it is to list the applications for which those principles have been used. Once instrument design becomes based on a set of basic rules and facts rather than be a further adaptation of what, perhaps, some relatively inexperienced person found could be used, then instrument realization will progress, carrying scientific advance and industrial success with it.

A hindrance to the rapid introduction of such ideas is the fact that instruments are generally designed only for small production quantities, making available only a low absolute profit in instrument design fields for any given product line.

A case study

An account of a specific instrument design project that can be considered to be systematic and based on sound theoretical foundations now follows. It is presented to suggest the approach that should be adopted, to a lesser or greater extent as the circumstances allow. The completed instrument is reported in Hugill (1984).

Project and Resources — Definition

The project arose as a doctoral research topic that primarily sought to apply the latest methodology in order to establish and improve systematic design procedures for instrumentation. It was decided that a suitable method for testing current approaches was to design and test an actual instrument. The instrument selected was a gravity meter that can be used to measure gravity variations down the length of a deep bore-hole. This was chosen as a research medium because it combined mechanical, electronic, and thermal regimes into a single instrument system and because in the fields of geophysical exploration such an instrument is not yet adequately developed. A gravity meter provides measurement data on the force of gravitational attraction existing between an object and the Earth.

The designer had majored in physics, with postgraduate work in geophysical studies. He had also attended a leading school where instrument systems design was a specialty. The project was supervised by a person with interests in rigorous and systematic design of instruments. Throughout the design period, progress notes were made of the methodology. Periodic reports were issued in order to keep track of the administrative design processes as well as for the methodology. Technical assistance was available to manufacture mechanical components and to create electronic systems.

In the approach taken, management for design quality and experience gained carried more weight than in a commercially 'management-by-objective' situation where time and cost are generally significant. This form of definition is a vital first requirement to establish in any instrument design task.

Establishing the Requirement

Actual manufacture and use of components is made possible after specifications are generated that define them and their configuration. In the systematic method of design, the specification-raising stage is almost the first one; the majority of the design effort goes into deciding what is needed before any manufacture begins. At the very beginning the designer must, however, first learn about the requirement so that suitable apparatus can be designed. Many authors have presented flow charts of the design method. Each portrays the same basic procedure. Those by Finkelstein (1977) and Bosman (1978) are shown in Figure 8.4. Further accounts are to be found in Asimow (1962), Beakley and Chilton (1973), Gregory (1968), and Marshall (1978) — some of the many works on design methodology.

Several key forms of activity must take place to establish the requirement. These are generally conducted in parallel to allow for delays in the procurement of information. A look at each activity follows.

Review Current and Past Designs

These can be studied from the published literature, including manufacturers' data sheets, and from experiences of the design team. It is especially important in mechanical design to give recognition to work done in the past. Unfortunately, many libraries, including the very best, tend to store texts and journals before around 1950 in very inaccessible places. This

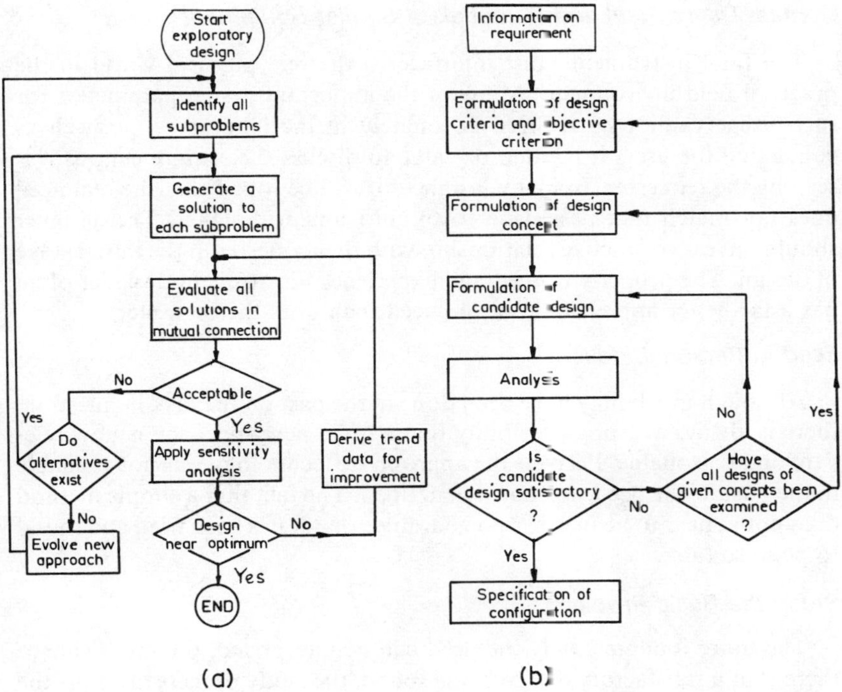

Figure 8.4 Flow charts on the instrument design process; (a) as given by Finkelstein (1977); (b) as given by Bosman (1973) (© Journal of Scientific Instruments. Used with permission.)

barrier to study of older works must not be allowed to prevent such a search for, as this text has shown, much of the mechanical design that might be useful is often to be found described in past accounts. Many computer-based data files contain only post–1970's information.

When reviewing a design, it is expedient to create specific scientific guidelines to follow. These must evaluate scientifically based parameters that can be compared in a scientific manner. This generally means establishing the physical principle involved and its characteristics. For example, in the gravity meter study each design was studied in terms of its static sensitivity and its sensitivity to influence parameters, this being almost entirely from geometrical considerations. Many of the pubished designs reviewed did not give this information (it probably had not been done!), requiring the geometrical equations for each to be generated. In many ways, relying upon quoted performance figures as a measure of a reputed instrument's performance can be misleading and often of little use in comparisons at this level of the design study.

Contact Users, Real and Potential, of Similar Equipment

The final instrument must interface to the real physical world in the practical field environment. Many of the important design parameters for such usage cannot be realized adequately in the laboratory. As well as contacting the users it is often of value to discuss the requirement as it is seen by the requester. In many instances it will be found that the real need does not match that passed on to the instrument designer. The designer should have a very active relationship with the requester in the initial stage of design. The properly trained and experienced instrument designer often has a far better appreciation of the need than does the requester.

Feed in Personal Ideas

Although much may have been done in the past on the task in question, there is always a strong possibility that a quite new approach is now relevant and reasonable. Perhaps the appropriate technology was not available in the past or the need was not well defined. The fact that a simple method does not appear to be in use is no guarantee that it has been tried and found to be unsuitable.

Study the Basic Problem

The more fundamentally the need can be interpreted, the more chance there is of a satisfactory design being found; the study process prompts the mind and provides the material for later tuning. A balance must be struck, however, between time devoted to the design task and that available to create what is needed. For example, in the gravity meter study it was deemed impossible to examine critically every design that had to be reported or could be conceived. A time limit was decided in which many designs were appraised. Every design team has this decision to make. It is the very reason for actual design objects finally being made and also for there always being the possibility that a more attractive alternative may yet be implemented.

Study Analogous Situations

Although not made for the same application, there is often another field of endeavor available in which similar tasks are undertaken but for different end reasons. In the case study it was found that the theory and practice of weighing microbalance design was relevant to gravity meter development. Another example, from another study, was that where the development of position-sensitive photocells was being undertaken in nuclear energy research groups without knowledge of the work taking place in general optics institutions; the devices were almost identical.

After six months of study (the project was envisioned to be a three or four year program of reported designs and new possibilities, which brought out the important parameters of a design for a bore-hole gravity meter, a report was compiled (Hugill, 1978) in that case as the need for a seminar. The report became an important review document for the design team as it recorded where the thinking had reached and for what designs. (It is this area that a computer prompt backup system would have been useful, as considerable detail had been studied that had become more and more difficult to retrieve as time passed.)

Good record keeping is vital. It obviously aids recovery of fact, but it also proves to the designer that progress has been made. It is human to rapidly accept new knowledge gained by study and not to be able to feel that progress has indeed been made.

This chapter describes, in a general manner, how to undertake a design. To make such a description logical it is necessary to make it appear here as though progress is made in a singular path. In reality, the designer is constantly updating information, often returning to past aspects. This is necessary because design conclusions can only be made when adequate quantities of information are available. Often to get to that stage requires the designer to make assumptions that are to be checked and verified later as information becomes available. Furthermore, any attempt to follow the course from generalities to specifics in a singular path will be extremely inefficient. For example, obtaining a certain literature article might take six weeks or more. It is necessary to form a progress route that allows for such long time constants and yet advances design with as much time devoted to the project as is allowed. Upon occasion it often seems, at the moment, that a certain factor is so important that it must be resolved before progress can occur. Abnormal effort is often poured into this aspect in order to break the nexus. This situation must be guarded against; in many instances at the early stage of a design study it is quite possible to circumvent the problem due to the multiple design possibilities then existing.

From this general study stage will emerge the specifications of what the instrument is intended to do. These are not the specifications of the configuration or of how to build the instrument, but to describe how it should perform. They may be a little unrealistic and they may be found ultimately to be incomplete. They are, in fact, often constantly tuned as the design proceeds (where that is possible, for in some cases these will be given to the supplier as the specifications to match or exceed).

These specifications must be as honest as possible and as realistic as is thought practical. They form the basis for the following stages of design and, therefore, must be well researched. In some very large organizations, staff are employed to develop these from within the company; they know the need better than the supplier. Where different agencies write and manufacture to a common specification set, it is important that the two parties arrive at the specification together.

In the case study it was found, by such means as utility analysis (see Figure 8.5 from Hugill, 1978) that a bore-hole gravity meter must be able to operate in small diameter bore holes, work properly at temperatures in the

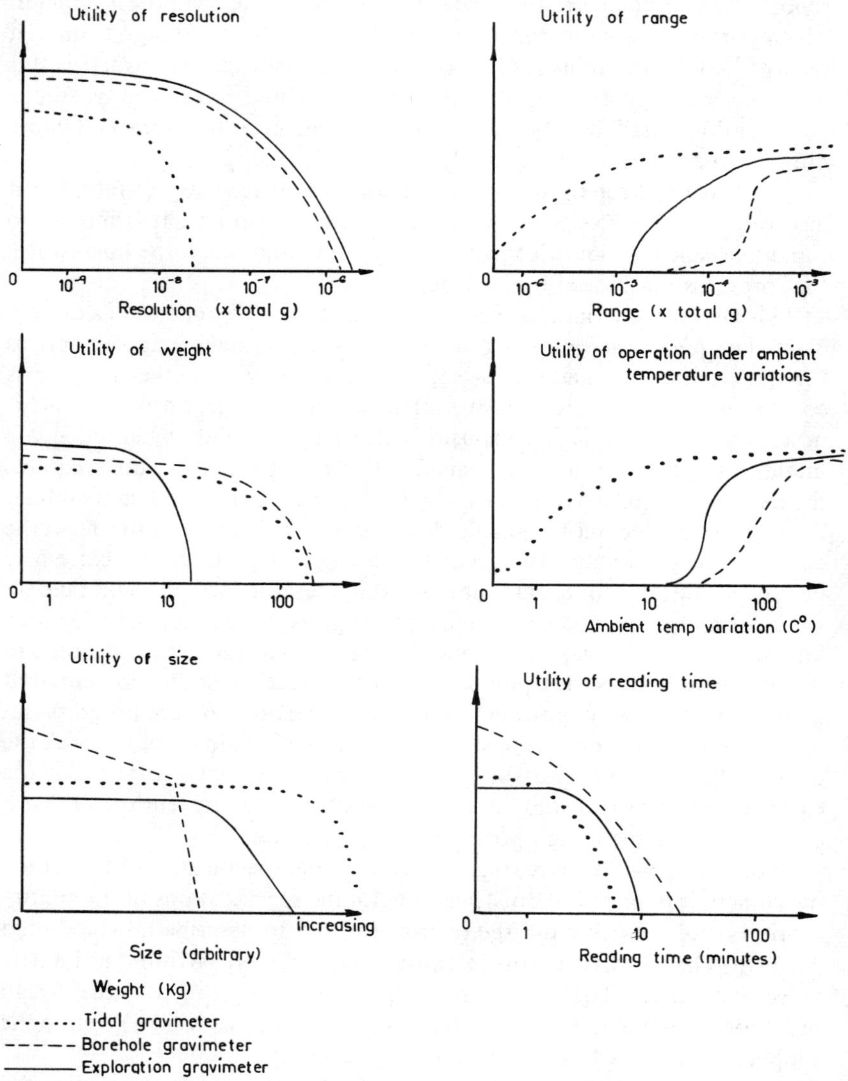

Figure 8.5 Utility of design characteristics assist the generation of specifications, Hugill (1978)

hole from ambient to 140° C, be able to operate from remote controls, and operate when tilted from the vertical by as much as 15 to 20 angular degrees. These are all additional requirements beyond those of the conventional surface exploration meter. As few instruments were in use, it was necessary to study the literature of temperature profiles in bore holes, the technology of the lining of holes, ways of aligning the meter to arc-seconds by remote control, and means to control the temperature at these elevated levels. Temperature levels alone mean that almost all currently used surface gravity meters are quite unsuitable. Temperature, being an important parameter, meant that study was needed of designs that had not been considered in this way previously.

From the information compiled it was possible to construct a total system block diagram (Figure 8.6) that portrayed the relationships between important parameters such as the gravity meter, the enclosure, and the environmental functions. The diagram shows their interactive relationships; it well demonstrates that an instrument is not an isolated piece of apparatus when placed in use. This form of diagram is a most valuable statement of the designer's situation and, although not always generated, should be given a high priority in the many tasks needed in the design process.

Selection of Candidate Designs

Having laid down the initial requirement specifications and having, by that stage, become very familiar with a wide variation of design possibilities at a very fundamental level, the designer reaches the stage where certain designs emerge that apparently (but see later) are suitable for further study. In order to keep actual hardware manufacture options open as wide as possible at this early stage, designs are chosen on the basis of a class of principal use, not on the actual design of manufacturers' construction.

Around this stage many designers tend to opt for an actual apparatus and begin full-scale manufacture of prototypes. This is not taking a systematic approach; the effort expended on such low level information can all

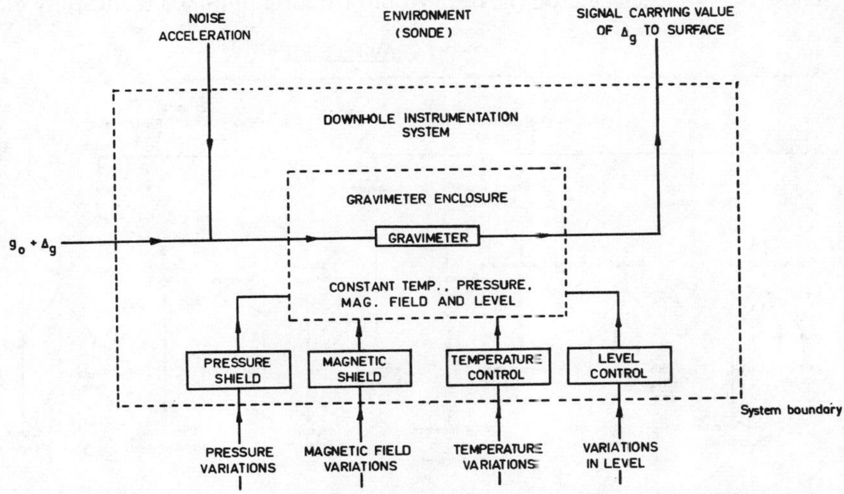

Figure 8.6 Block diagram of envisaged bore-hole gravity meter system in use (Hugill)

too easily commit the designer to following a rigid design path regardless of several design faults that may, and usually do, occur.

The value of using mathematical and physical part-systems models at early stages cannot be overstressed. These enable the design to be optimized, studied, and even abandoned if need be without serious loss of manufacturing effort. Preliminary designs are merely the beginning point for more detailed analysis of expected performance. In the case study the majority of the reported plus some other conceivable designs for gravity meters were tabulated as a tree diagram. Many forms of these diagrams are usually possible. The method considered useful in this instance was to order them in terms of the static and dynamic physical principle involved. The chart developed is shown in Figure 8.7. Another possibility could have been to order them in terms of their use for relative and absolute measurements, but that was less useful as a design aid. Such charts may need to be non-flat due to interconnections spanning the groups. Such levels of sophistication are generally avoided if possible.

At the extremities of the chart sit the many kinds of design being classified upon the principle used. In the initial stage they are not studied for relevance; that comes next. From the many possibilities, several designs are chosen that appear worthy of a more detailed inspection. At this stage it is inappropriate to settle hastily for any particular one to be developed because the level of the designer's understanding is still too limited.

In this design study it was clear from practical and basic theoretical considerations that the mechanical elastic suspension branch was most suited to the task of measurement in a bore hole. Within that branch the concepts most likely to succeed were narrowed down to five geometries. These five were selected on the basis of information obtained from study of

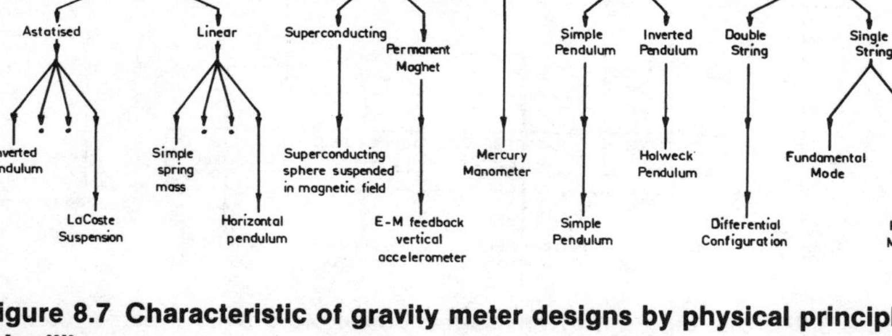

Figure 8.7 Characteristic of gravity meter designs by physical principle (Hugill)

archetype mathematical models of the several design concepts embodied in that branch.

An idea of the depth of theory needed to make a reasonable evaluation is shown by the following extract about the single inverted pendulum from Hugill (1978).

> Consider the pendulum shown in Figure 8.8. The pendulum rotates about an axis through 0, and is acted upon by a gravitational torque T_g and a spring torque T_s.
> The two torques operating about 0 are
>
> $$T_g = mgh \sin \theta$$
> $$\doteq mgh\, \theta, \text{ small } \theta$$
> $$T_s = \tau\, (\theta + \theta_0),$$
>
> where θ_0 is some initial rotation of the spring. At equilibrium $mgh\,\theta = \tau\,(\theta + \theta_0)$ (as $\tau > mgh$ for stability, $\theta_0 < 0$) which can be solved for θ
>
> $$\theta = \frac{\tau\theta_0}{mgh - \tau}$$
>
> Differentiating this expression with respect to g gives
>
> $$\frac{\partial\theta}{\lambda} = \frac{-mh\tau\theta_0}{(mgh - \tau)^2}$$
>
> The mechanical sensitivity can be increased as mgh approaches τ. The advantage of this system over the horizontal pendulum is that it does not rely upon a high ratio of mh/τ to increase mechanical sensitivity. Hence, it is potentially more compact and robust. The chief disadvantage is that it is extremely sensitive to tilt; another disadvantage is that the sensitivity is highly dependent upon g.

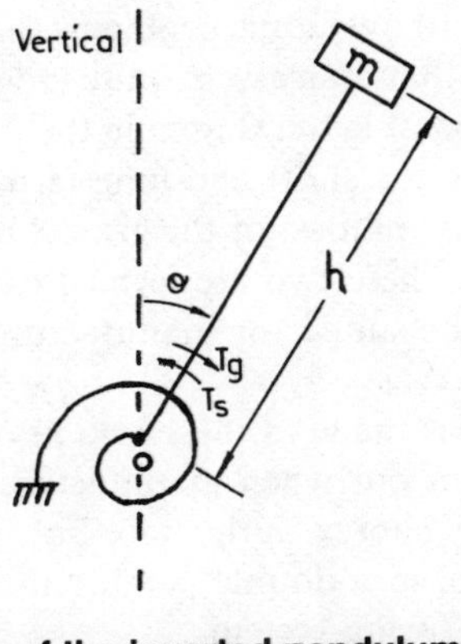

Figure 8.8 Schematic of the inverted pendulum gravity meter, Hugill (1978)

It is not always necessary to develop mathematical models in order to rule out a plausible concept. Practical limitations of a technological nature often make it clear that certain designs are not applicable in a given application. For example, the gas spring concept was ruled out because changes to gas pressure caused by the large temperature variation that will be experienced in a bore-hole would be hard to eliminate. The vibrating string concept was known to be less than reliable enough due to the high stresses placed upon the string to obtain adequate performance. A 1940's patented idea, in which gravitational force is converted into friction via a rotating bearing and thus temperature rise (measured as a variable) related to gravitational force change, was thought to be quite unworkable at the sensitivity needed.

Thus the original infinity of ideas was systematically brought down to five possibilities. The next step was to review these in a quantitative manner in order to reduce them to a smaller number. It is, however, quite possible that all may be found suitable, in which case further work is necessary to optimize the choice.

The five selected were analyzed in greater detail for their sensitivity to gravity input, for their ease of setting of the zero, and for their range of operation. Interesting features of existing meters in use emerged at this stage. One instrument, for example, possesses very high sensitivity but the price paid (analytical design revealed this) was the need for an expensive tight tolerance zero setting arrangement. A second example, drawn from another instrument study about a form of special mechanical spring being sold on the market, was found from analysis to be using at least 50% more material than was needed for the performance sold. Furthermore, the more optimum spring dimensions would also provide a smaller, lighter spring with improved dynamic response.

It became clear from several different case studies that, in general, many instrument designs reported and sold have been raised by nonsystematic methods in which the design has virtually stumbled along with periodic injections of rearrangement to overcome problems that have been found in use. It also became obvious that many were quite independent reinventions and that a greater proportion of better design in the first stages would have been highly advantageous in the short and long terms.

From the five selected possibilities for the gravity meter, it became clear that two might be suitable. These two are termed candidate designs; they are designs that appear to be suited for manufacture as the final specific product.

Further detailed study was made of these two in order to establish any feature that would rule them out when manufacture and application are considered. The two initially chosen in the bore-hole gravity meter project were the inverted pendulum, in a double pendulum form (a new innovation), and a buckling beam spring system.

These two were again analyzed at greater depth than any of the previously considered units. Sensitivity analyses were conducted on the mathematical and experimental subsystem models. It was subsequently found that for the inverted pendulum design the two pendulums must have their axes of rotation parallel to within 2.5 10^{-8} rads to yield adequate sensitivity. This was deemed to be impractical. The other design was also ruled out as it was established that the required residual flatness of the spring material used to form a buckled beam spring element exceeded normal supply expectations.

Had either of these been built without these detailed sensitivity studies, many months of work would have been largely wasted before practical tests would have eventually revealed these major shortcomings as unexplained defects of design that had not been properly identified.

With both candidate designs ruled out, it was necessary to go back (illustrating the value of good records and detailed work) to the chart and select a new possible candidate design option. A form of horizontal pendulum having elastic pivots was selected. It has not been used in gravity meters before but some of the features are used in seismometers; hence, the need to watch for analogous situations.

Detailed sensitivity analysis showed this design to be worth pursuing and at some stage it was decided that it was reasonably safe to commit this candidate design to the actual constructional stage.

So far, in the case study being described here, design has not concentrated on the details of actual construction except where it has seen that such factors may effect the choice of principle adopted. Choosing the right principle is a most important step. As an illustrative example adopted from another instrument design project (with some alterations to highlight the points of interest), consider the measurement of length using a tensioned wire drawn from a measuring drum. One unit developed, apparently with little theoretical understanding, began using a small electric motor to provide the tension needed via a very large gear ratio. This restricted the speed at which the wire could be withdrawn or fed out. To increase the paying out speed, (the withdrawal rate was left the same), the wire was allowed to extend faster than the tensioning motor system, which then needed time to catch up with the wire before a measurement could be indicated when the calibrated tension was reached. The design inherently included backlash between the measuring drum part of the system and the tensioning, motor-driven mechanical digital readout. To overcome this defect a direction-sensing mechanical switch was used to insert a fixed offset change in the reading. The complete unit was continually being redesigned as the gearing and fine mechanics of the system were themselves not designed using basic theory. The project reached, through sheer doggedness on the part of its developer, the stage where a few prototypes were made. Marketing began on the strength of the best of the results obtained. A well known firm purchased the rights to manufacture. They, however, very rapidly discovered that it was not a repeatable design in terms of low-cost mass production and the project was shelved.

In contrast, another far more satisfactory project with the same design aim began by studying the statics and dynamics of a tensioned wire being tensioned by a hypothetical torque motor having the basic parameters that would be involved, that is, rotational inertia, bearing friction, limits of speed, and the like. This led to the creation of a system in which the wire was attached to a measuring drum, which was tensioned by a special (Negator type) spring torque motor. The torque motor was designed for minimum inertia so as to provide fastest dynamic response. The bearing system was designed for adequately low static friction in order to achieve good slow-to-static speed measurement fidelity. The net result was a system having no gears in the measurement chain, far less parts than that described above, negligible backlash, and very much faster feed rate; at least two orders of ten improvement were obtained in the measurement precision and accuracy. The manufacturing cost of the unit was estimated to be considerably less than for the previously described unit. In essence, using the right principle can make or break a design's prospects.

Once a prime candidate design has been selected, the design, at least until proven otherwise, assumes some degree of rigidity. A principle, however, can be implemented in a variety of ways. The designer's next task, therefore, is to work systematically toward a specific construction, again using rigorous procedures wherever possible.

In order to pave the way towards hardware decisions, it will possibly be necessary to consider several systems aspects since these, where they apply, may influence the choice of such parameters as layout, materials, and interconnections. In the case study the choice existed to use either the horizontal pendulum in the open-loop measurement mode (where deflection of the pendulum is a direct indication of the gravitational change from the datum zero position) or to implement it as a null balance device (in which the beam is forced back to a datum null point measuring the parameter value, producing this as the indication). Each method has its particular advantages. It was thought from a study of the factors for each that a null balance would be used. This decision had an important bearing on the overall physical shape and architecture of the system and the space needed.

It also then became feasible to establish which parts of the total system could be considered as stand-alone subsystems, units that could be designed and tested largely in isolation. This was assisted by the creation of a further block diagram that detailed the system lying within the enclosure shown in Figure 8.6. The use of icons is highly recommended as they provide the designer and those with which he needs to communicate with a succinct portrayal of the featues of the system that are of particular interest. The importance of pictorial representation in technological development is described in Ferguson (1977). Electronic subsystems are more easily modified than are mechanical modules, and for this reason electronic systems tend to be assembled largely through the trial and error design method. However, it cannot be stressed enough that the use of systematic methods based upon theoretical and pseudo-theoretical knowledge (empirical equations, for instance, deduced from experimental data) will usually lead to a far more satisfactory result.

Although it may be possible to separate out stand-alone subsystems modules in a design, it must not be forgotten that they ultimately must be interfaced into the complete system. For this reason they must be developed, when done in isolation, with due regard to their total system functional requirements. Such factors as the matching impedance, rate of response, power capability, and joining arrangements may need careful attention. In electronic systems many of these parameters have been designed out of the user's area of control by the use of additional circuitry (output buffering, input protection, voltage stabilization, for example) but in mechanical systems the state of the art does not include the means to adopt such measures, leaving it very much to the user to ensure that compatibility results.

Subsystems modules can also be tested for their tolerance to influence parameters. In the gravity meter case study, it was apparent that the sensor first chosen to measure the beam's displacement might be influenced by the magnetic fields known to exist in steel-lined bore-holes. A special test rig was established to investigate this. Again, theoretical considerations of magnetic fields and of shielding were investigated as part of the practical testing program of this aspect.

In like manner the various subsystems were each studied at the general theoretical level to the point where meaningful and obsolutely essential construction was needed either to advance the designer's knowledge or where it was clearly possible to invest manufacturing time in a part that had a very high probability of being used in the final system.

Construction Design

Eventually the stage is reached where at least some of the total system can be manufactured. Where a large design team works on the project, many subsystem parts will be developing at the same time. In the smaller project, where only one or two persons are involved, it will be their shortage of time that will restrict progress. It will be periodically necessary to set priorities. In many cases a number of subsystem modules will have equal priority, in which case no special order is needed. This was the circumstance in the case study once the mechanical aspect of the beam and sensor were frozen on concept.

The selection of the specific geometry and sizing of the spring, mass, and sensor were found from the study of the sensitivity charts made previously, from the experience gained with test of methods by which the delicate parts could be formed, from information gained (after some considerable delay) about wire availability, and from other fundamental data that had become available from the previous, more fundamental investigations.

It is at this stage that the beneficial use of systematic procedures of designs begins to make itself apparent. For example, it was eventually learned that the only available size of the low thermoelastic coefficient metal wire on the market was slightly different from that used in the optimization calculations. As it was not possible for reasons of finance or time to have a special batch made, the design of the spring was adjusted, using charts and equations so as to yield the same sensitivity in the same overall space. Even without building the unit, the designer has, at this early stage, a high degree of confidence that it will perform closely to the requirements. Compare this with the more usual method of iteration through a number of prototypes toward a reasonably satisfactory one in which other subsystems are tailored to make up for the deficiencies of the basic initial unit used as the primary key.

The theoretical studies also brought out several interesting factors that would not have been so readily realized from a purely experimental development method. For example, the lowest thermal error results at a given temperature of operation of the metal. As each batch of wire has a different postmelt history and slightly different composition, it became clear that the optimum instrument must be one in which the jacket control temperature is adjusted to suit the actual material used. Unfortunately, the available theory of this effect in the metal is not sufficiently developed for the best temperature to be accurately predicted before use. This was another facet of the design that could not economically respond to the use of as rigorous methodology as would be preferred. (A boundary of knowledge was reached in that respect and it was not appropriate to set up a research program to attempt to extend it.)

During the initial construction stages, special production methods were carefully pursued if it was clear that they would be needed in the manufacture of the final instrument. The shaping of the coil spring, the flexure pivots, and a method of repeatable assembly of the unit were examples in the case study. This kind of information becomes very relevant again when, and if, the instrument goes into production.

Throughout this explanation the use of systematic methods based upon theory has been stressed. It must, however, be said that a balance is needed between the academic approach and the purely practical alternative. A balance must be struck between 'management to objective' and 'management for perfection of understanding'. It is not too difficult to find that enormous time has been consumed developing very elegant theories for the design of a part that has low priority for inclusion. Systematic procedures will usually pay off more handsomely in the long term than the ad hoc alternative although, at the time, they may be less rewarding in so far that a specific design detail does not become a working part so rapidly.

Testing of Basic Principles and Elimination of the Unexpected

As subsystems are designed they should be tested as far as possible in readiness for inclusion in the whole. The practice of first making all parts and assembling the whole without prior part testing is not generally an efficient means of progressing prototype development work. The predominant reason for this statement is that there are too many chances of malfunction in development. The whole system becomes far too extensive to faultfind and rectify when so many factors are interacting. Pretesting of subunits helps to overcome this and also provides increased confidence that the whole will operate satisfactorily when put together. (A developed and tried production design can be assembled ready for acceptance testing; in that case there are unlikely to be many, if any, defects if the production processes are in control.)

Emphasis on systematic design and good theoretical understanding does not entirely guarantee that the system will operate as planned when first assembled. This is because design is based upon certain prior knowledge from which the specifications of use and of structure are gleaned. As the designer does not start with adequate and total knowledge of the situation, there is always a considerable chance that some assumptions that had to be made will prove to be incorrect. The role of testing, especially at the subsystem stage, is to verify that the unit does perform as hoped.

When testing shows up an unexpected defect, the design must be altered in some way to eliminate it. Beware of tacking on modifications to correct for lack of performance caused by incorrect basic design. The description of the wire length measuring device, given above, illustrates the kind of situation that can develop. Such modifications often end up controlling the flexibility of the designer's choices, if not virtually freezing all movement in design flexibility. It is good practice to eliminate defects in development apparatus by as much basic redesign as is needed. At that stage the cost of alteration is far less than it will be in a production program.

As procedures of testing are themselves subject to likely error, it is important to establish the validity of tests and to take accurate notes of the procedures and results of the tests. As this is part of the learning process about the design, any error in this stage can negate correct information with possible gross implications. At the time of testing, this information often seems to be of little relevance to later work, for the test is being conducted to establish if a subsystem is acceptable. If, however, the testing of units made later, such as occurs in the first preproduction batch, shows up a defect, it will become very important to see if the tests were wrong (those later tend to be made in a state of less ignorance), or whether the design is wrong. If it is the latter, then the whole production cycle must be broken. If the tests were found to be wrong, then the design may still be acceptable.

It is, therefore, important that all prototype units be retained in the same condition as that in which they were first tested. If improvements are needed as the result of these first subsystems tests, it is more prudent to make new subunits than to modify the existing one. This practice enables more accurate assessment of the performance of the unit as it was first built and accepted. It will assist any preproduction runs to be compared with these former models.

When sufficient modules are made and tested, they can be combined, usually as one additional unit at a time to keep the number of problem sites to a minimum, in order that higher level tests can be made to establish if defects are introduced by the interactions created.

It is not possible to state exactly how any particular system should be developed; that is where experience and the designer's ability become important. However the testing and assembly are carried out, the main goal is to maintain control of sources of problems, because as soon as too many simultaneous sites occur it becomes far more difficult to retain an orderly, systematic development procedure in which confidence in the design is maintained.

To illustrate the above points, consider some steps of the case study. It was clear that the mechanical spring mass balance subunit would predominate the general design of the gravity meter in terms of the size, shape, and design rigidity of the various subsystems involved. For these reasons it was the first part developed as a stand-alone unit. Several full and part models were made at approximately the correct size. These provided the testing and learning ground for tuning the design of the mechanical regime.

They were not made with any degree of exactitude as far as appearance or size constraints were concerned. Only those features that need precision were included at the correct level of manufacturing finish. By relaxing some of the total manufacturing tolerances in this way, it is easier to reach the point of interest; this is acceptable provided the relaxed parameters do not influence the results sought at that time.

In many cases it may be possible to physically simulate the mechanism of interest using one of the several available commercial mechanical construction kits. These are offered over a range of building precisions, one being capable (see Figure 8.9) of providing assemblies having runouts and deflections in the order of 10–20 μm.

Where the exact material of components may not yet be available, it is often possible for a temporary substitute to be pressed into service. It must not, however, become a key factor unless it can properly be regarded as a replacement in the final design.

In the gravity meter project the early spring-mass balance was sensed with a manually observed microscope. This enables sensitivity and frictional errors to be estimated as well as an eventual electric sensor could.

Figure 8.9 Sample of the complexity to which constructional systems may be used to simulate a final mechanism, courtesy FAC System AB, Gävle, Sweden

When moving on to develop the sensor, the crude but evaluated balance then became the test ground for various facets of the sensor because it could be considered as a calibrated force-measuring device.

Field Testing

Many instruments are made for use in an application that does not apply the same laboratory conditions as those in which they are developed. As the instrument, in such cases, must perform properly under possibly more stringent and hazardous conditions and where influence quantities may be far different from those experienced in the laboratory, it is usually necessary to conduct special tests that simulate (or better still, attempt to totally emulate) real field conditions.

It is not always convenient, or efficient, to directly take the instrument from the laboratory to its final environment to make these tests. Instead, evaluations are made of the design's performance using special test apparatus. For example, temperature, humidity, and pressure influence effects are conveniently studied in an environmental test chamber in which these parameters can be varied in a systematic manner while the instrument is being observed and is in operation. Other tests prior to application might include those for mechanical strength and life of electrical connection points, or to seek information on the noise pickup level of acoustic or radio frequency electromagnetic radiations.

It often is necessary to construct special test facilities; in some projects the testing facility can cost more to construct than the instrument, a cost that is well justified in certain circumstances. In the case study, influence of possible variations in the Earth's magnetic field on the sensor that was used required construction of a Helmholz coil arrangement that was used in a vibration-free, temperature stable laboratory. Figure 8.10 shows the apparatus. Its lack of manufacturing care is defended as being unnecessary for the application.

A factor often overlooked is the need to ascertain if the electrical, pneumatic, hydraulic, or whatever power supply is employed in the field does indeed provide power that is adequate in terms of the power level available, the frequency spectrum of its ripple, the stability of its variable, and the

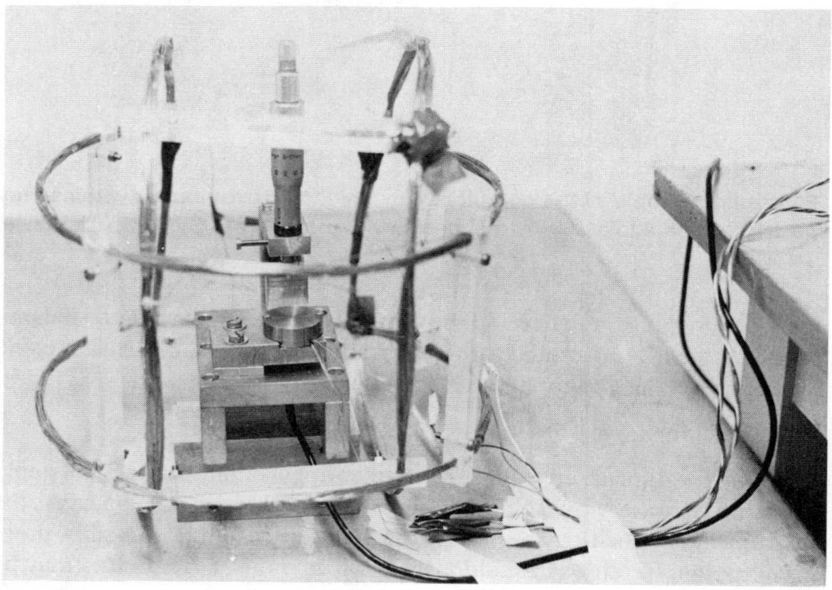

Figure 8.10 Special test apparatus may be needed to evaluate the whole, or part of a prototype, see text (Hugill)

suitability of its matching arrangements. Tolerance to such variations can be investigated in the laboratory using specially devised tests.

Tests made in the laboratory can never totally emulate the reality of the actual application. An instrument thoroughly evaluated for one application should be reevaluated for another. One set of circumstances may not be a duplicate of the other. It is not always a weakness in the designer's skill if unexpected shortcomings reveal themselves in the final use of an instrument. This is so because the designer never constructs a system with absolute and total knowledge of the field circumstances. No matter how much effort is expended, there will always be a finite chance that a defective condition can occur. However, it is possible to greatly reduce the chances of this occurring by employing an experienced designer and by giving greater attention to the condition under which the instrument will finally be used; in short, by expending greater effort on the testing and specification raising stages. Aircraft instrumentation, for example, has its extremely high reliability and usefulness because of quite vast experience in testing and design, coupled with long experience of carefully policed use and practices. At the other end of the cost spectrum, a hastily constructed instrument made on the spur of the moment to satisfy the immediate need of a research progam may prove to be quite inadequate.

Many examples exist of instances where instruments have failed to operate properly when placed into final service. There were to the writer's knowledge, at least two magnetometers made that worked in the local region but that failed to operate at other locations because their design had not allowed for the different ambient magnetic field levels. A very large and costly seismic marine survey failed because it was discovered, after the towing ship had recorded data for many days, that the vibration of the cable connecting the hydrophones to the ship introduced noise into the signals.

The degree to which a prototype should be tested depends upon the importance of the task of the instrument. Unfortunately, it is not easy to assess the amount of testing needed to fulfil this criterion, and management is generally reluctant to allow enough testing for cost or time reasons. Clearly, however, a Moonprobe must operate on the first time of use as no second or third chance exists. Similarly, a balloon-borne gamma-ray telescope needs to be well tested since only one 24 h flight results from each two years of preparation.

Extrapolation of the cost versus reliability and correctness of operation characteristic might suggest that at the other end to the Moonprobe levels of reliability it is not necessary to conduct tests at all, making use of actual field data as the means to test and finally tune the design. This is taking the theme too far. Under field conditions it will often be found that more than one defect results. In the field it will be difficult to identify the source of a problem, since it cannot easily be isolated. Furthermore, design alterations made in the field are often far from properly engineered.

Where feasible, the procedure should be to simulate the field conditions known to exist in an orderly, systematic manner, rectifying the design accordingly as the tests proceed. Then would follow a field trials phase in which the operator and designer would use the instrument in a practice run. This will highlight defects. An example of the vagaries of problems is a solarimeter program in which carefully coated and shaped glass domes collected radiation from the sun, measuring the received level with photocells, which was held up because spiders continually spun webs in the field of view. Blowflies have been known to collide with fine mechanical suspensions, breaking them.

Field testing is never really finished; the unexpected can occur at any time. A rainfall recorder station placed in a remote dry bush location recorded no rain for two years, only to be completely washed away by the first rain. As was once remarked to the writer by a student engaged in placing strain meters to measure earth rock strain deep in old mine workings, the hardest part of the project was to actually achieve a satisfactory day's work in the field. There is always the possibility that some small defect — a lost part, a forgotten tool, the weather conditions, and lack of necessary skill on the site — would hold up installation for yet another week.

The process of evaluating an instrument design for use in a given application has been given some order by the laboratory practice called "evaluation". Certain laboratories, such as the Sira Institute in Britain, conduct such tests on behalf of a client. This practice is, however, only in its infancy and will continue so until management recognizes the real financial cost of a poorly made measurement. At the time of writing, the gravity meter of the case study had not reached field trial stages, so it was not possible to give examples of field trials from that study. It is felt, however, that the other examples presented here have sufficed to illustrate the difficulties that must be faced and the possible procedures that can be followed in order to bring an idea about an instrument into a real, actual, physical object that will serve the measurement purpose for which it was built.

It seems highly appropriate here, as this account is brought to a close, to reiterate the philosophy once expressed succinctly by a leading academic studying the systematic design of instruments. He said that designers and users of instruments should adopt the attitude that they think all instruments are no damn good. If this attitude is adopted, the user will constantly be on guard for defective data.

And then to production

This chapter has only commented upon the methodology employed to bring a measuring instrument design from the initial stage to one where an apparently satisfactory prototype results.

If that instrument were intended to be made in larger numbers, such as for sale as a product line or for use in large numbers as part of a large experiment, then the new phase would be construction and evaluation of a larger batch, say, of five to ten units. Tests of these will invariably reveal more faults that should be rectified by fundamental changes to the design (if that is still possible). Where the small batch is the precursor of a much larger product run, called the preproduction batch, changes to the design should be more carefully done, for the production instrument will be very costly to alter after it is made due to the large numbers made and the problems of recall from customers with the resultant lack of goodwill.

This chapter does not attempt to describe how a prototype is taken from that stage into production. It is, however, important that the reader realize most positively that the problems of production can be quite different from those of prototype construction and that taking an instrument from the tested prototype stage to the marketplace is a far more costly and time consuming process than designing the prototype.

For example, a firm wishing to market a simple, reasonably straightforward design that is closely like an already marketed product can reach the prototype stage from the decision of what they intend to market in a mere two months. To take it from there to the point where sales can commence with adequate level of after-sale service, prior promotion, and stocks issue may need two years.

Typical differences having impact are the methodology of component manufacture (such as the use of die casting of parts instead of the more costly one-off manufacture with machine tools), the use of production-run ready-to-use components, and special batch orders. Streamlining of the design for low-cost manufacture and assembly and alteration of the design to allow for easier servicing to accommodate extension into further product lines and the use of options — those, plus many more facets, are generally not in the province of the prototype designer's realm if production is not envisioned. The importance of having conducted sensitivity analyses of the system becomes very apparent at this juncture. As has been expressed earlier, it is vital that factors important to production be incorporated into the various stages of the prototype design since they are far more easily accommodated at that stage. Any instrument design that might reach production should involve the skills of a production engineer from the very beginning.

References

Asimow, M. 1962. *Introduction to design*, Prentice-Hall, Englewood Cliffs, NJ.

Beakley, G. C., and Chilton, E. G. 1973. *Introduction to engineering design and graphics*, Macmillan, New York.

Bosman, D. 1978. 'Systematic design of instrumentation systems', *J Phys E: Sci Instrum*, **11,** 97–105.

Ferguson, E. S. 1977. 'The mind's eye: nonverbal thought in technology', *Science*, **197,** 4306, 827–36.

Finkelstein, L. 1977. 'Instrument Science: Introductory article', *J. Phys E: Sci Instrum*, **10,** 566–572.

Gregory, S. A. 1968. *The design method*, Butterworths, London.

Hugill, A. L. 1978. *Gravimeter design*, City University, London, Internal Report, Systems Science, DSS/AH/171.

Hugill, A. L. 1984, 'Design of a gravity meter', PhD. thesis, Flinders University, South Australia.

Marshall, R. C. 1978. 'Ideas to order', *Electronics and Power*, January, 54–57.

Whitehead, T. N. 1934. *The design and use of instruments and accurate mechanism*, Dover, New York.

INDEX

About the Author

Peter Sydenham has been Professor and Head of the School of Electronic Engineering, South Australia Institute of Technology, since 1979. He holds an ongoing Visiting Research Fellowship at UMIST, UK, and a Visiting Professorship at the City University, London.

Dr. Sydenham is a member of the IMEKO TC-1 Higher Education Committee, President of the Institute of Measurement and Control, Australia, a Director of the Australian Scientific Industry Association, and a Fellow of the Institute of Measurement and Control, London.

He is Editor-in-Chief of the new John Wiley series on *Measurement Science and Technology* and has written or contributed to eleven books. Most of his some fifty published primary papers are in his specialist area — computer-aided design of instruments — and in establishing systematic measurement science